SLINGSHOT SPECTACULAR

THE FRONT-ENGINE DRAGSTER ERA

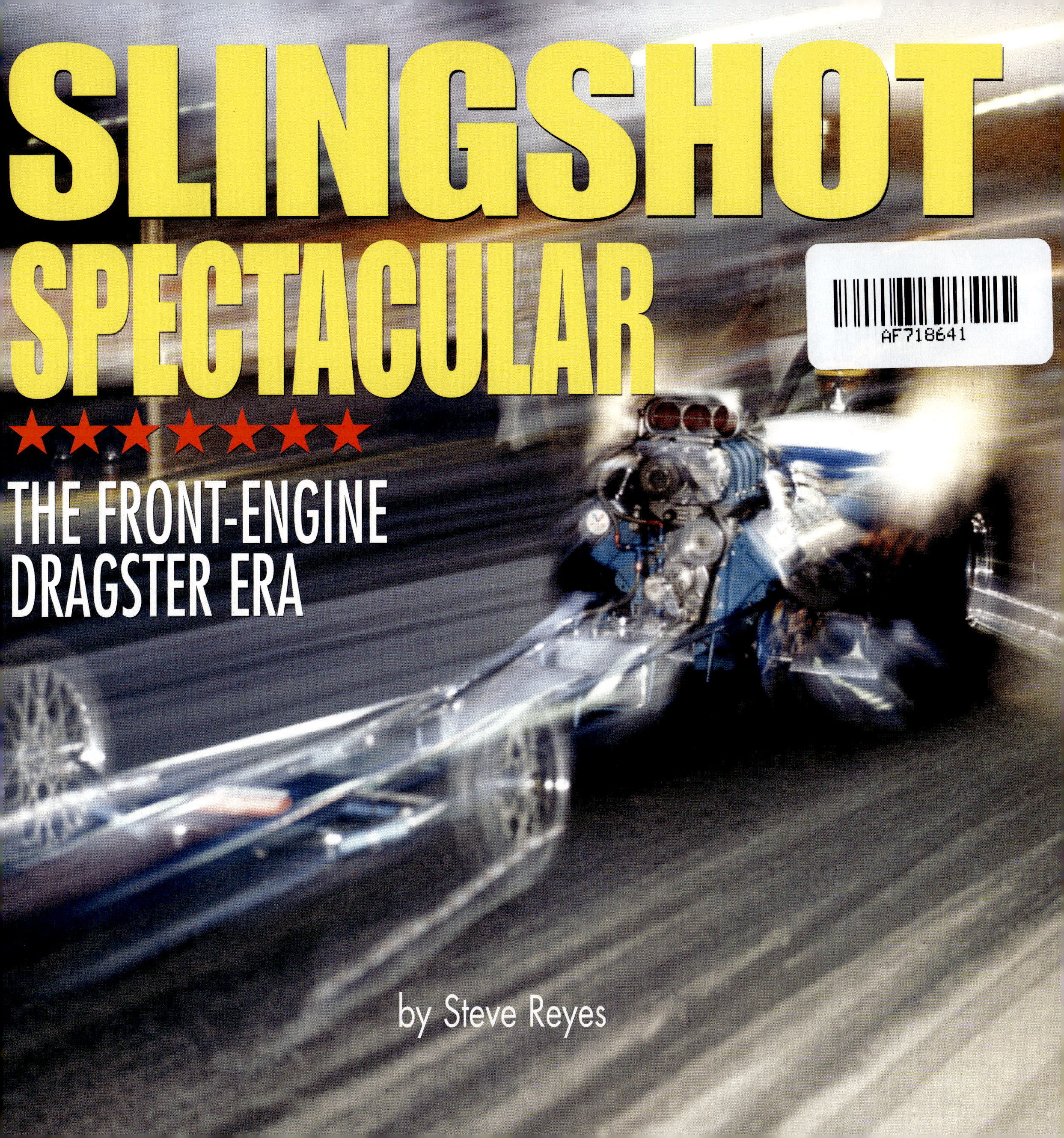

by Steve Reyes

Edited by Travis Thompson

Designed by Connie Nordrum

ISBN-13 978-1-932494-64-8

Printed in China

CarTech®
39966 Grand Avenue
North Branch, MN 55056
Telephone: (651) 277-1200 or (800) 551-4754
Fax: (651) 277-1203
www.cartechbooks.com

Library of Congress Cataloging-in-Publication Data

Reyes, Steve.
Slingshot spectacular : front engine dragster era / By Steve Reyes.
p. cm.
Includes index.
ISBN 978-1-932494-64-8
[1. Automobiles, Racing.] I. Title.

TL236.R476 2007
796.72--dc22

2007027754

Cover:
From Plains, Virginia, came the husband-and-wife team of Jim and Alison Lee with Tom Raley at the controls. Alison was the first female crew chief of a Top Fuel dragster. In September 1970, their Great Expectations *Top Fuel car set the NHRA E.T. record at 6.63 at Atco Raceway in Atco, New Jersey.*

Title Page:
Captain Billy Tidwell in the Kuhl & Tidwell Top Fuel car at Orange County International Raceway in 1970.

Back Cover Top:
Tony "The Loner" Nancy blasts off from the starting line with his Wynn's Sizzler *Top Fuel car at Lions Drag Strip in Long Beach, California.*

Back Cover Middle:
Danny Ongais carefully stages Carl Casper's Young American *Top Fuel car at Scottsdale, Arizona in 1970.*

Back Flap:
Don Moody gets ready to launch the Cerny-Lins Top Fuel car at Irwindale in 1969.

OVERSEAS DISTRIBUTION BY:

Brooklands Books Ltd.
P.O. Box 146, Cobham, Surrey, KT11 1LG, England
Telephone 01932 865051 • Fax 01932 868803
www.brooklands-books.com

Brooklands Books Aus.
3/37-39 Green Street, Banksmeadow, NSW 2109, Australia
Telephone 2 9695 7055 • Fax 2 9695 7355

Contents

Dedication

For Eric Medlen 1973–2007

Acknowledgments

My files of *Drag Racing USA*, *Drag News*, *Drag World*, *Drag Sport*, and *National DRAGSTER* and of course, all those men and women who put on a fire suit and attacked the quarter-mile. Thank you!

Introduction

It was 1963 when I thumbed through the sports pages of my dad's *Oakland Tribune* newspaper, and an ad caught my eye. It read, "Come See the Nitro-Powered Slingshot Dragsters!" For a wide-eyed 15-year-old, nitro-guzzling race cars sounded like a very interesting and fun-to-watch proposition. I didn't have a way to get to Fremont Drag Strip, which was 25 miles away. Begging and pleading, I finally persuaded my dad to take me.

Some racers just plain scared me!

Fremont Drag Strip had smells and noises I had never experienced, and the sweet smell of nitro filled my nose and lungs. Hearing the sound of a nitro-burning engine was like traveling to a different dimension. I was quickly addicted and needed to see, hear, and smell Top Fuel dragsters. Starting out, my nitro lifestyle wasn't easy, as the patience of my "taxi" (dad) wore very thin very quickly. Then I obtained a Honda 305 Dream motorcycle that gave me the freedom to explore the great drag strips of Northern California.

Ron and Jody Lawrence managed Fremont Drag Strip,

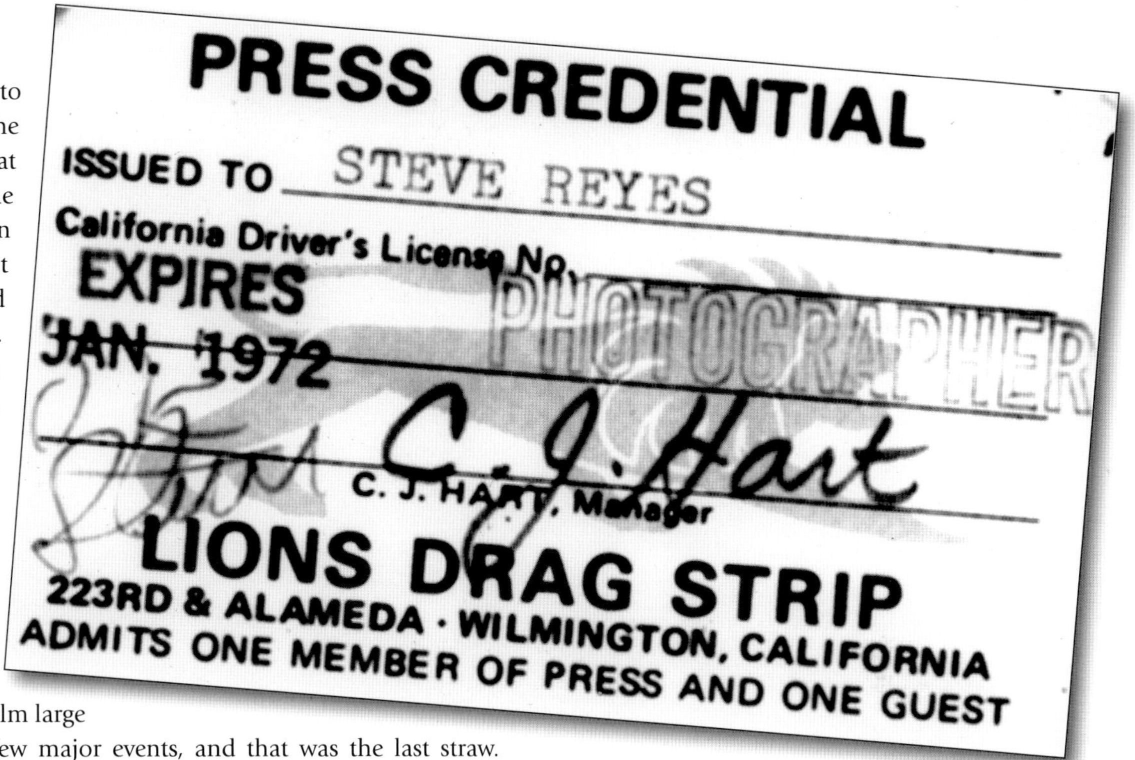

and they gave me my first chance to capture Top Fuel cars on film. The film I refer to is movie film, as at first I thought I was going to be the George Lucas of drag racing. I soon decided that movies were okay, but I wanted to stop the action and show motion in still photographs. My longtime friends, the Jackson brothers, Jamie and Sonny, filled the bill for the drag movies in NorCal. So armed with my 8-mm Eumig movie camera, a 620 Kodak Brownie, and a wind-up motor-drive Instamatic, I prepared conquer the racing world. Although the Lawrences permitted me to film in Fremont's photo areas, I was not allowed to film large or special events, so I missed a few major events, and that was the last straw. Finding a way to shoot at every race was my goal. Back then there were few other photographers, and they didn't want this wide-eyed kid getting in their way, so I didn't get much help from them.

Picking up the latest drag papers of the day, I figured the best way to have access to all events was to get published in all drag-racing publications. Goodbye movie camera, hello 35-mm camera. Purchasing the drag-racing newspapers of the day helped me develop my eye for a good photo. I studied the photos of Jim Kelly, Tim Marshall, Mike Mitchell, Jere Alhadeff, and Bob Hardee. To me, they were artists who painted with film and cameras. Drag strips were my stomping grounds between 1963 and 1973.

It seemed like anybody who was anybody in drag racing came to California to race. From October to March the drag strips on the West Coast roared with the sounds of Western, Midwest, and East Coast Top Fuel racers. All I had to do was wait and the best drag racers of the day came to me. All the big names of the quarter-mile were moving in front of my lens. Like a kid in a candy store, I rejoiced from weekend to weekend. I relocated to Los Angeles in 1970, when it seemed like drag racing's heart was pumping in Southern California. There were chassis builders, engine builders, custom painters, and of course race tracks, and I now lived in the center of it all. I was five miles from Keith Black, 10 miles from Lions, and 20 miles from Irwindale and OCIR (Orange County International Raceway).

The neighborhood I moved into included Jack Chrisman two streets over. In fact, the duplex I moved into had belonged to Bill Shrewsbury, the *L.A. Dart* wheelstanding king. Moving from Northern California to Southern California was like having nitro mainlined into my veins — 24 hours a day, seven days a week, drag racing. I was photographing race cars, developing film, printing photos, and visiting all the drag-racing publications based in Los Angeles. It was a crazy lifestyle, but lots of fun, fun, fun!

THE RISE AND FALL OF THE SLINGSHOT

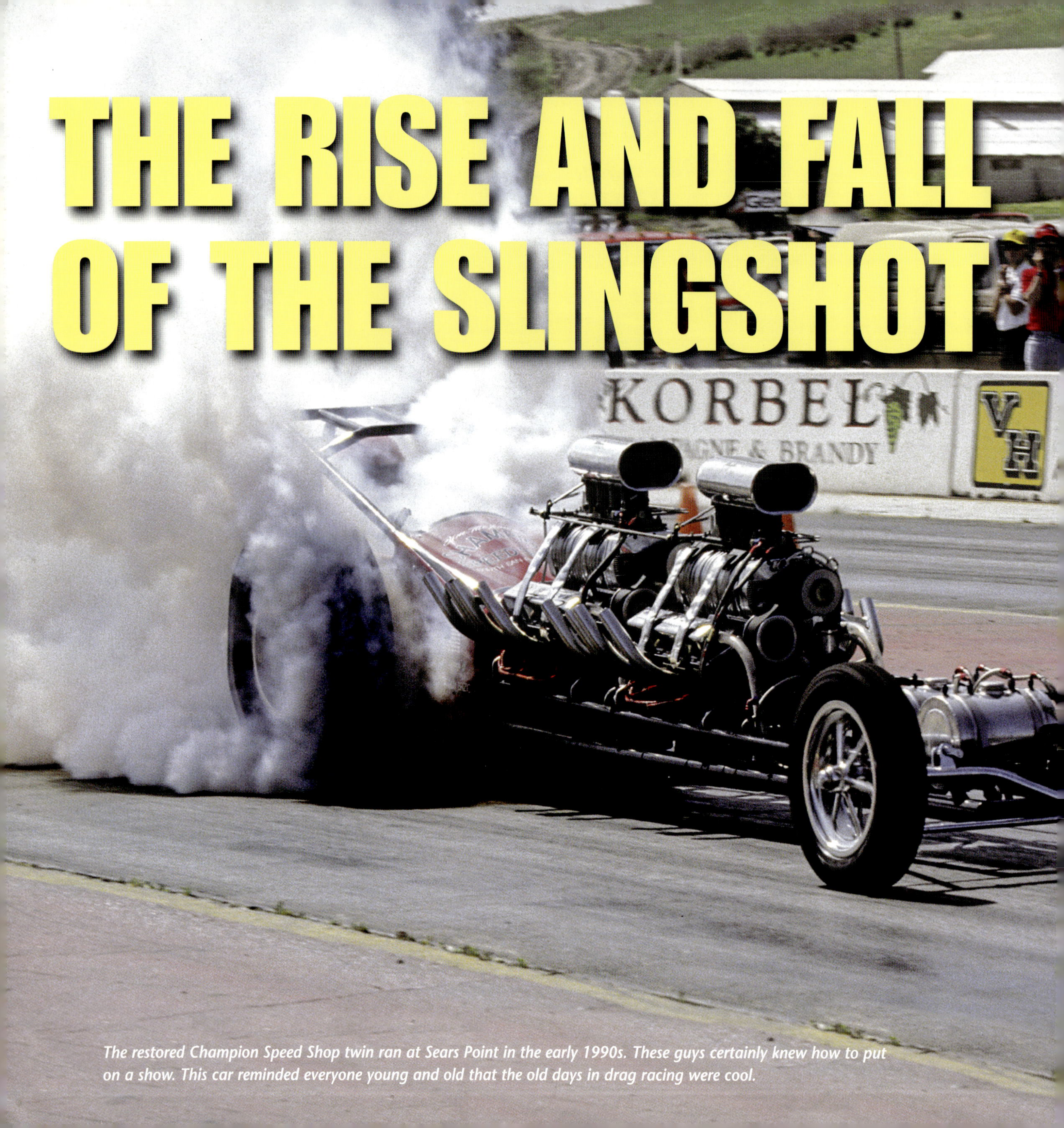

The restored Champion Speed Shop twin ran at Sears Point in the early 1990s. These guys certainly knew how to put on a show. This car reminded everyone young and old that the old days in drag racing were cool.

Nineteen-forty-eight was a good year, at least from my perspective. I was born and drag racing held its first organized drag race at Goleta, California. The Santa Barbara Timing Association hosted this historical event. From there it seemed like drag racing grew very quickly, and the fabulous 1950s started with good ol' C.J. Hart, Frank Stillwell, and Creighton Hunter negotiating a deal with Bill Nichols. Nichols was the operator of the Santa Ana Airport, and Hart and friends wanted to use an abandoned taxi strip for weekend drag races. The first race at Santa Ana was held July 1, 1950. By the end of 1950, three drag strips were running in Southern California, including Santa Ana, Pomona, and Paradise Mesa near San Diego.

To get racers off the city streets, Wally Parks started the Southern California Timing Association which in 1951 became the National Hot Rod Association. Classes were created according to the cars of the day based on engine and tire size, fuel type, and primarily speed. Body style was not a consideration. The dragster class included the fastest cars or motorcycles. Calvin Rice won the Inaugural Championships in 1950.

A dragster was a basic stripped race car that consisted of a lightweight or modified body to fit the chassis consisting of an engine, a fuel tank, and a modified lightweight front end. Sometimes the body and the frame had holes punched or drilled into them to make the car lighter and faster. These dragsters were also very dangerous as the adjustments could make the chassis weak and a crash fatal. Early on the roadster was king of the early drag wars but the dragster, or rail, changed the sport forever. A couple of racers have received credit for building the first actual dragster. Dick Kraft was the first to run a bare-frame car on July 3, 1950, when he blazed the quarter-mile at 113.92 mph in the *Bug*. The "other" first dragster, was a flathead-powered set of '29 Chevy rails built by Art Chrisman and Holly Hedrich. So you can take your pick as to who was first. The 1950s just kept churning along for the sport of drag racing!

In 1951, Drag City opened, and Fontana, which is 60 miles east of Los Angeles, was the place to race. This is the same Drag City that inspired Jan and Dean to write the hit *Drag City*. Fontana was to become well known for Top Fuel shows as well as being the only track to run four jet cars down the strip at one time. Needless to say, the four jet cars running at once was only attempted one time. The wind and turbulence behind the jets tore up the starting line, tower signs, etc. The drag strip at Saugus, California, also opened its gates for the first time in 1951. Saugus was quite a tow from a major metropolitan area, but was still able to draw well-known racers of the time.

DRAG WORLD

July 2, 1965

TWENTY-FIVE CENTS

VOLUME 1, No. 18

FEATURE INDEX

GARLITS RETIRES

The first time I saw Don Garlits was at Fremont in 1965. He had just beaten everyone at Bakersfield and was looking to do it at Fremont as well. Alas, he went out in the semis at Fremont. A couple of months later Drag World declared, "Garlits Retires." A crash back east with the Frantic Four and the deaths of Tex Randall and Lou Cangelose in their Fuelers prompted Garlits' decision. His retirement didn't last long, and soon "Big Daddy" was back to kicking everyone's butt in the Top Fuel class.

Drag racing was starting to get the attention of fans in Southern California in 1950–51, and when nitromethane fuel was adopted by California drag racers in 1952, the sweet smell of nitro filled the air and its popularity exploded. For those of you who don't know, nitromethane isn't really fuel but an explosive with a low stoichiometric number so you can jam a lot of it and oxygen into a cylinder. There are eight cylinders, each with its own explosion, that propel the car down the track at record speeds. For instance, a 9-second car would go into the low 8s with a proper amount of nitro. However, for the lower ET you bought yourself a lot more broken parts.

Also in 1952, Mickey Thompson, the late "King of Speed," made his presence known racing the first blown fuel Chrysler dragster. It was a simple dragster with a wedge motor and homebuilt rails. In 1953, Art Chrisman and Leroy Newmayer hit a top speed of 140 mph. Joaquin Arnett drove the *Bean Bandit* dragster to 134 mph at Paradise Mesa Drag Strip. Up in Northern California, Kingdon Drag Strip opened in Lodi.

There was no slowing down the drag racers of the 1950s; the fuse had been lit and the sky was the limit in this new high-horsepowe sport. Mickey Thompson again made it known he was

In a typical mid-1960s battle for Top Eliminator at Fremont, the Bailey Brothers **Dragmaster** *(far side) took on John Skinner—blower versus injectors, Hemi versus Chevy and the Bailey Brothers went home with the win.*

an innovator and top racer when he debuted the sport's first fully covered dragster, called the *Panorama City Special*, in 1954. Thompson didn't stop with that landmark in drag racing; he went on to introduce the slingshot dragster design, which incorporated the narrowed rear-end principle. This dragster had the driver sitting over the narrowed rear-end housing, with about two-thirds of the car close to the rear wheels for traction control. The fuel tank and weights were up front to hopefully keep the front wheels on the ground. You will notice that the cars often were named the drivers' name or nickname, especially for those who were on a budget and didn't have the funds for the extra details such as paint schemes and lettering.

We don't want to leave out the Midwest drag racers. In 1954, the first

Jim Kelly is the best photographer ever to capture drag racing on film. The easygoing Southern California native set the standard for drag-race photographers. Kelly and I toured a couple of times and had lots of fun. However, once we got to the races, Kelly was all business, and his drag-racing images are classics. In 2003 he was inducted into the International Drag Racing Hall of Fame, one of four photographers to have achieved that honor.

While on tour in the early 1990s, I came across **The Dolphin**, *and owner Dick Bertolucci kindly let me shoot a couple of rolls of film of the restored 1954 dragster.* **The Dolphin** *had raced in Northern California during the 1950s, but when his business took off, Bertolucci parked the racecar. After it sat for more than 30 years, Bertolucci restored his racer and displayed it at car shows in the NorCal area. He wanted the young kids to see what he raced when he was a wild and crazy drag racer.*

World Series of Drag Racing was held in Lawrenceville, Illinois. The highlight of the event was Art Arfons racing at 132 mph in an Allison-engined dragster. The Allison engine was a powerplant from a WWII fighter plane and was readily available as government surplus after the war. Back in the West, Ollie Morris and his rear-engine flathead ran 145 mph at Santa Ana. This was very fast for an ol' flathead. Rear-engine dragsters could be seen throughout the United States, but there were problems with their handling and with getting consistent speeds and ETs.

The late Terry Ervin's **Junkyard Special** *was a longtime Fremont, California, fixture. When Ervin's ancient car couldn't compete in Top Fuel, he just removed the blower and ran in C/FD.*

The mid-1950s brought a new drag strip to Southern California—Lions Drag Strip in Long Beach, and the new manager was none other than Mickey Thompson. He knew how to market drag racing to the fans, and this track definitely made a splash with the spectators. The lanes were close together with the spectator stands right up against the track to help fans feel like part of the action.

Milestones in 1955 included the first 150-mph run on gasoline when Lloyd Scott cranked out 151.01 in the dual-engine *Bustle Bomb*. Then that guy Thompson ran 151.26 mph in his blown fuel Chrysler. Right behind Thompson was Chrisman running 145 mph in his blown Chrysler.

Rudy Babino's **Velvet Hammer** *was outdated in 1965, but Babino still tried to compete in the Northern California Top Fuel wars. The pretty purple digger was a Fremont regular in the mid 1960s.*

In January 1971, the Wildlife race team of Don "The Snake" Prudhomme and Tom "The Mongoose" McEwen debuted their new "square-bodied" Top Fuel cars. I spent a fun day photographing the Mattel people doing a television spot on the diggers. The Top Fuel cars never did run the way Prudhomme and McEwen wanted them to. The only win I remember Prudhomme's car getting was at the 1971 Northern Nationals at Fremont. Ironically, Prudhomme put Mike Snively in the car for this event, and Snively beat Rick Ramsey and the California Charger *in the final. McEwen's digger was sold to Van Iderstine's Speed Shop. Art Marshall drove the car, and it was the last front-engine fueler to win an NHRA national event. Sadly, Marshall was killed when he crashed the car at an East Coast track. Prudhomme's white car is now on display at Don Garlits' Museum of Drag Racing in Ocala, Florida.*

The Los Angeles area loved its drag racing, and this was proven when a crowd of more than 10,000 saw the 1955 Southern California Championships at Pomona. NHRA also made its mark with the first NHRA National Championship in Great Bend, Kansas. Rounding out the drag-racing action in 1955 was Calvin Rice defeating Fritz Voigt to win the Arizona National Championships with a thundering 10.30 at 141.95 mph in his Top Eliminator dragster. Rice continued his winning ways in 1956 when he won the first Bakersfield Championships with his Chrysler-powered dragster at 156.46 mph.

The birth of the American Hot Rod Association (AHRA) came about in 1956, and the AHRA held its championship event in Great Bend, Kansas. Then San Gabriel, California, opened on May 6, 1956, and on the first day, Chrisman ran 142.73 mph with a 10.24-second ET, but then Ollie Morris ran for Top Eliminator against Voigt and won with a 10.12-second run. Actually, the overall win went to Tom Augers and his nitro-burning Vincent motorcycle when he clocked 10.80 at 129.68 mph. In the meantime in central Florida, one racer was wreaking havoc in Top Fuel and Top Eliminator racing. Don Garlits and his *Swamp Rat I* took top honors at the Florida State Championships with a 10.90 at 135 mph.

Linda Vaughn was the queen of drag racing accompanied by her Hurstettes, Marcia (left) and Niki. I first met Linda in 1966 at the* Hot Rod Magazine *race. She had just become Miss Hurst Golden Shifter.

In 1957, NHRA banned nitro because racers had not yet perfected its use and were blowing up, catching on fire, crashing and exploding. This wasn't very popular in California because racers there had been putting a little "pop" in their fuel tanks since 1952. So if you raced at an NHRA-sanctioned event, you raced gas only. Just prior to the nitro ban, Cook and Bedwell blasted down the quarter-mile in 8.89 seconds at 168.85 mph with their nitro-burning Chrysler. The famed *Speed Sport* roadster of Greth, Fisher, and Maynard made its presence known in 1957 by cranking out a pass at 169.11 mph, and Garlits' ol' buddy Setto Postoian posted a victory at the World Series of Drag Racing with his 9.36, which was the best of the event.

A couple of drag-racing highlights from 1957 included "Jazzy Jim" Nelson's dual-flat-

On September 19, 1965, Emory Cook brought the* Jocko's Invasion *streamliner to Fremont. The Allison-powered liner made a few unimpressive runs and was put in the trailer, never to return to the Fremont quarter-mile.

At the Half Moon Bay California Championships in 1965, Don "The Snake" Prudhomme squared off with NorCal's favorite son, Denny Milani, in the first round of Top Fuel. Milani left too soon and lit the red light, giving Prudhomme and the **Hawaiian** an easy victory.

I think the team of Leong, Black, and Prudhomme won everything in 1965, and the California Championships at Half Moon Bay was another notch in their belts. The beautiful **Hawaiian** Top Fuel car waded through a stellar field to claim the grand prize.

head digger running 9.60 seconds at 146 mph at Saugus, Emory Cook winning Top Eliminator at the AHRA National Championships with a 9.35-second ET in his Chrysler, and Art Arfons' *Baloney Slicer* hitting 166.97 mph at Columbus, Ohio. The *Baloney Slicer* got its name from the prop left hanging from its Army surplus V-12 Ranger engine. It was definitely a happening year.

In 1958 that Florida guy, Garlits, was at it again with his *Swamp Rat I*, the first dragster to run faster than 180 mph and the first to use bicycle front wheels. Phil Weiand supplied the first eight-carburetor manifold to the *Swamp Rat I*. Garlits ran an 11.80 with the 353-ci Buick-powered dragster at the NHRA Nationals.

The Italians were able to enjoy drag racing for the first time at Pomigliano Airport in Naples, Italy. Setto Postoian again took Top Eliminator honors at the World Series of Drag Racing with 9.06 ET

at Cordova, Illinois. Back in California, "TV Tommy" Ivo took time off from his television career to crank out an all-time top speed on gas of 159.01 mph. Bobby "The Beard" Langley set the pace at the Bakersfield Championships with 9.07 at 166.0 mph in *The Scorpion*. At the same Bakersfield track, Tony Waters sped to 8.99 at 174.41 mph in his A/M roadster called the *Barnstormer*. Nitro ban? What nitro ban?

There were good and bad moments in 1959 drag racing. On the West Coast, Jim Deist introduced the sport's first drag chute for dragsters. This helped save cars and lives. On June 20, 1959, during a run at Chester, South Carolina, Garlits' engine exploded and his face and hands were badly burned. He spent the next five weeks in a hospital. Art Malone took over the driving chores for Garlits until he could again get behind the wheel of the *Swamp Rat I*.

Back on the West Coast, Art Chrisman and the Chrisman/Cannon *Hustler* were winning in Bakersfield with a speed of 178.10 mph, and then 181.91 mph for another win. The other Chrisman in drag racing, Jack, drove the *Sidewinder* Chrysler of Mailliand, Chrisman, and Jones

"Slammin' Sammy" Hale has been around forever. He drove for Ted Gotelli and a number of other race teams in the 1960s. Here he is at the wheel of the Champion Speed Shops AA/FD. These photos were taken at Sears Point Raceway in Sonoma, California, in the early 1990s.

During late 1964 this blown Olds dragster made its debut at Fremont Drag Strip. I don't remember seeing it ever again. It got lost in the drag racing abyss.

Drag racers will build just about anything to try going fast, and this dragster proves the theory. This unknown entry didn't set the record books on fire, but it sure was different.

to a 9.04 ET at Lions. Finally, 1959 ended on a down note. Drag racers on the West Coast lost a favorite place to race when Santa Ana closed and was turned back into an airstrip.

The 1960s bloomed, and so did the sport of drag racing. The Aussies got their first taste of drag racing when Castlereach Drag strip in Sydney hosted the first Australian drag race. The race was attended by 5,000 fans who watched 12.5-second elapsed times by 60 entries. Meanwhile, back in the United States, Ivo made news again when he piloted his twin-engine Buick to an 8.95 ET on gas, the first 8-second run at Pomona. The unlikely duo of NASCAR and NHRA joined forces to bring the world the first Winternationals in Daytona Beach, Florida. Top Eliminator was won by Lewis Carden with a 9.55 ET.

Garlits was still on the mend, so Art Malone took the *Swamp Rat I* to the West Coast and on March 20, 1960, defeated Ivo to win the Northern California Championships at Fremont. Art "The Dart" blazed to an 8.42 at 187.10 mph. In July 1960, Garlits was back in the saddle of the

Swamp Rat I. Then to round out 1960, Leonard Harris won the NHRA Nationals in Detroit driving the Albertson Oldsmobile.

Garlits continued to be troubled in 1961. During testing at Golden Triangle eighth-mile drag strip in Oldsman, Florida, he experienced another bad explosion and suffered more burns. This time Connie Swingle took over for Garlits, and Swingle wasted no time in *Swamp Rat III*. He smoked to an unofficial ET of 7.88 seconds at Columbia, South Carolina. The first NHRA-only-sanctioned Winternationals was held at Pomona, and Hayden Proffitt set top speed of 176 mph in a twin-Chevy digger. To close out 1961, Jack Chrisman was named Drag Racing Driver of the Year by *Motorlife Magazine*.

The Belmont Boys, Walt and Jesse Schank, drove a blown Desoto with Algon injectors on a 97-in chassis in the early 1960s. On August 12, 1961, they set the standard quarter-mile B/FD speed record at Lions Drag Strip at 183.28 mph.

About 100 years of drag-racing photographic experience are represented in this photo. Four of the best ever to capture drag racing on film are, from left, Richard Brady, Bob McClurg, Jere Alhadeff, and Tim Marshall. The still photographers were the unsung heroes of the sport. When television and other mainstream media ignored drag racing, these guys kept it in the public eye with great images of drag racers and the sport they loved.

*The **Glass Slipper** was restored and ran at Sears Point and Bakersfield, California, in the 1990s. It was originally constructed in 1954 by Ed and Roy Cortopassi and Doug Butler of Sacramento, California. It was credited with being the first dragster to combine a streamlined fiberglass body, aluminum frame, and enclosed canopy. Dick Butler drove the restored **Glass Slipper** in the early 1990s and now it rests in the NHRA Motorsports Museum.*

With Dave Zeuschel power, Ivo started 1962 by making the West Coast's first 7-second run, a 7.99 at San Gabriel. The man from Engle Cams, Don Moody, cranked out a 7.76 ET at Pomona in the Moody and Davis Top Fuel car. Jack Chrisman lowered the gas mark to 8.34 at San Gabriel while driving Mickey Thompson's Pontiac digger.

After almost an 18-month layoff, Garlits returned. At the NHRA U.S. Nationals, he took the wheel of his friend Connie Kalitta's Top Gas dragster to run for a national record of 180.36 mph. That year, Garlits also received his nickname, "Big Daddy," when he kept winning race after race. If you thought Garlits sat around while healing from his injuries, you would be wrong. He and his crew churned out 25 chassis for their customers.

Happy days returned in 1963—I discovered the world of drag racing *and* the NHRA lifted the ban on nitro! The NHRA decided to allow nitro again because there were much better fire suits

"Wild Bill" Alexander, was still doin' it at the California Hot Rod Reunion in Bakersfield in the early 1990s after a retirement of about 20 years. He was able to return to his first love of slingshot Top Fuel dragsters and hadn't missed a beat. He can still be seen wheeling a Top Fuel car at NHRA Hot Rod Reunion events in Bakersfield, California.

available to protect the racers. With nitro in the tank and a wing mounted on his dragster, Don Garlits plowed through the Top Fuel field at the NHRA Winternationals, winning with an 8.26 at 186.32 mph. The 1963 Winternationals also marked NHRA's introduction of Top Fuel as a professional racing category. This was an unlimited class where you ran what you brought as long as you met the safety rules for fire suit, weight, etc.

Early in 1963, Don Prudhomme took the Greer, Black, & Prudhomme Top Fuel car to a 7.77-second pass at San Gabriel. Finally, the NHRA replaced the flag starter with the Christmas tree, or "matchmaker," starting system at the U.S. Nationals. ABC Wide World of Sports gave drag racing its first national television event coverage at that race.

B&M introduced its Tork Master automatic-transmission glide system for dragsters in 1964, and Goodyear Tire and Rubber Company entered the drag-racing tire field, introducing competitive slicks. It was interesting how the big winner of the week often dictated what slicks would be on all the cars the following week. It was fun to guess who would be running what tires from week to week. This was also a booming year for Don Garlits. He became the first driver to run 200 mph officially when he clocked a 200.44, then he won the NHRA U.S. Nationals with a national record of 7.67 seconds at 198.22 mph in his Wynn's Jammer *Swamp Rat VIB* against the late "Fat Jack" Williams in the final. Williams was driving the Crossley, Williams, and Swan Top Fuel car. In October, Garlits loaded his digger onto the *S.S. United States* and sailed for Surrey, England, with the rest of the U.S. drag-racing team. The team toured several drag strips, and Garlits set the British land speed record at 197 mph.

While 1964 was good to the *Swamp Rat*, 1965 was to be the year of the snake. Don "The Snake" Prudhomme opened the season by winning the NHRA Winternationals in Roland Leong's *Hawaiian* AA/FD, after which he went to NorCal and won the California Championships. It seemed The Snake was striking whenever he slid into Leong's *Hawaiian*. The one race Prudhomme didn't win was in Bakersfield, where Garlits served notice on the Top Fuel cars at the Smokers Meet. Garlits won everything and was runner-up to himself—sort of. Marvin Schwartz drove the Team Garlits No.-2 car against his boss in the final.

It was still Prudhomme's year as the Leong, Black, & Prudhomme *Hawaiian* won Top Fuel at the NHRA Nationals and was said to have pocketed $67,000 in winnings for the year. Although Garlits had won Bakersfield, he faced turmoil when he crashed twice, once in New Jersey and once in Maryland. Garlits retired again and Carl Schiefer began driving for him, but Garlits couldn't resist a match race of The Snake versus the *Swamp Rat* at Half Moon Bay, California. He dusted off

It was October 1966 with a jet versus a Top Fuel dragster at Fremont. John "The Black Knight" Edmunds driving Tony Waters' digger puts a holeshot on J.D. Zink and the Untouchable *jet car. "The Black Knight" put it to* Untouchable *three straight that night.*

In 1957, a kid actor from Burbank, California, blistered the quarter-mile in his Buick T-bucket with a 12 flat at 117 mph—that kid was "T.V. Tommy" Ivo. Not being content with those ETs and speeds, he partnered five years later with Dave Zeuschel and was able to make the first 7-second run on the West Coast, a 7.99 at old San Gabriel. "T.V. Tommy" was one of the true showmen of drag racing, from his beautiful show-and-go Top Fuel cars to his big smile and friendly demeanor, which made promoters and fans alike love him.

his fire suit and helmet and slid behind the wheel of *Swamp Rat X*. With Connie Swingle as his crew chief, he met Prudhomme for the showdown to see who really was No. 1. When the smoke cleared, the man from Florida and his sleek black fueler sent Prudhomme packing, and Garlits won the No. 1 spot.

Drag racing started off 1966 at Half Moon Bay, California, on January 23, 1966, where Garlits and Prudhomme mixed it up again. This time Prudhomme was at the wheel of the B&M Tork Master car, a full-bodied Woody car with Dave Zeuschel Hemi power. Garlits had his 426 Dodge Hemi-powered *Swamp Rat X* with its special water-pump cooling system. Over 10,000 people crowded into the drag strip by the Pacific Ocean and after two rounds of their three-match race, they were tied one each. The third race was a smokey side-by-side affair with Garlits doin' it to Prudhomme again. A deafening roar from the 10,000 fans assured Garlits he was truly King of the Dragsters.

If you haven't read or seen Jon Asher's work in the last 40 years, you must be living in a cave in Tibet. The outspoken Asher is a well-known writer and photographer, and his drag-racing knowledge spans over four decades. Of course, Jim Kelly and I taught him everything he knows about photography. At a rainout in Marion, Ohio, I showed Kelly and Asher how the game of eight ball is played. They still won't play pool with me.

Leland Kolb was long and lanky just like his racecars, and he was a fixture on the Southern California drag-race scene. Then Kolb was driving his 331-ci Chrysler-powered '32 roadster at Saugus, California, in 1954, setting the Top Gas speed at 125 mph; that was just the start of a great drag-racing odyssey for a class act. Kolb sometimes gave young raw talent such as Kelly Brown a chance to prove themselves in Top Fuel racing.

A familiar "oldie" was spotted at a Goodguys event at Sears Point in the early 1990s. The restored Charlie's Business Coupe *was out making laps down the NorCal quarter-mile. I also have photos of this car running in the mid 1960s.*

Two old smoke machines faced off at a Goodguys event at Sears Point in the early 1990s, the Champion Speed Shop twin versus Art Chrisman's Hustler I. *I don't recall who won, just that there was a lot of cool smoke and noise.*

With Prudhomme now driving for Bob Spar's B&M Tork Master car, this left the driver's seat vacant in 21-year-old Roland Leong's *Hawaiian* Top Fuel car. Two weeks before the 1966 NHRA Winternationals at Pomona, California, Mike Snively was called to fill the seat. Snively and Leong clicked and the *Hawaiian* won Top Fuel at the NHRA Winternationals two years in a row. Everyone then turned their attention to the March Meet in Bakersfield, California. In 1965 Garlits "owned" the place but 1966 would be different. Bakersfield was the largest gathering of nitro-burning dragsters in the United States, and the California trio of Skinner, Jobe and Sorokin (a.k.a. "The Surfers") Top Fuel team set a new low ET record at 7.34 and took home the top honors.

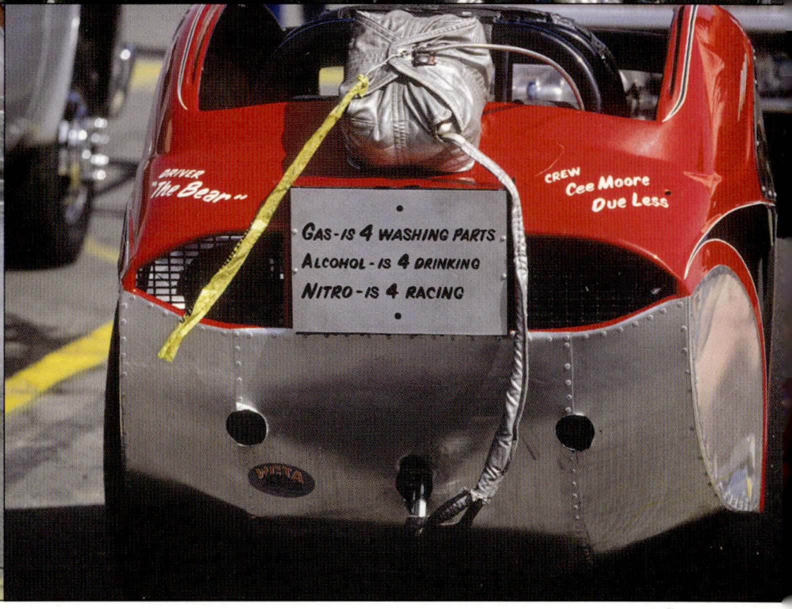

Restored and doin' what it did in the good old days, **Charlie's Business Coupe** *put on quite a show in the early 1990s at Sears Point. It still entertains its fans with lots of smoke and wheels in the air.*

This is a state-of-the-art Top Fuel car at Fremont, California, during the mid 1960s with an unknown owner/driver. This car was full–bodied with a padded rollbar, Zoomie headers and a 392 Chrysler Hemi for an engine.

Other cool stuff in 1966 included Connie Kalitta driving his 427 SOHC Ford Top Fuel car to the first accredited speed over 220 mph. He ran a scorching 221.12 mph for an ET of 7.35 at Capitol Raceway in Maryland. That guy Prudhomme was on tour and swept the first UDRA Memorial Day Olympics of Drag Racing three days straight with a best of 7.53 at 218.18 mph at the Rockford, Illinois, track.

Highlights on the West Coast included Tom "The Mongoose" McEwen driving the Baney-McEwen Yeakel Plymouth Top Fuel car to victory over Ron Rivero in the Frantic Four Top Fuel car at the Hot Rod Magazine Championships. Johnny "Black Knight" Edmunds drove the Waters and Edmunds

Top Fuel racers sometimes built rather strange cars to attack the quarter-mile. This unknown 1960s racer thought four-wheel drive and side-by-side injected engines were the answer for speed on the NorCal quarter-mile. Alas, they weren't.

Top Fuel car to AHRA Championship honors. Back on tour, Snively and Leong's *Hawaiian* won an unprecedented fourth NHRA National Top Fuel Title by beating out Danny Ongais at the 1966 NHRA Nationals. The year closed with a big surprise when John Mulligan drove the 392 Hemi-powered Adam's and Wayre Top Fuel car to a groundbreaking 6.95 at 221.12 mph at Carlsbad, California.

Connie Kalitta and his 173-in Logghe chassis 427 SOHC Ford Top Fuel car had a great year in 1967 when they won the 1967 NHRA Winternationals, the AHRA Winternationals, and the NASCAR Winternationals. This was a drag-racing Triple Crown win for the one-eyed driver/owner from Michigan. Nineteen-sixty-seven also saw the addition of a second car to the *Hawaiian* Top Fuel team when they were joined by a Don Long-built Top Fuel car driven by Mike Sorokin. At the Bakersfield, California, 1967 March Meet, the old *Hawaiian* with Snively at the wheel took Top Fuel honors. Soon after the victory at Bakersfield, Leong got a phone call from an Italian guy who wanted to buy the Kent Fuller *Hawaiian* and take it to Italy for the show-car circuit. They negotiated a price and the car was shipped sans its race engine because Roland put a Chrysler Marine 392 in the car before it shipped. Then Roland received a letter stating the car was donated to an automobile museum in Turin, Italy, and there it sits to this day. They later tried to get the car back, but it is in a private museum that does not allow selling or exchanging of items in the collection. So if you ever want to see the original *Hawaiian* you will need to pack your bags for Turin, Italy.

Prudhomme switched rides for the 1967 season to drive for Lou Baney in a 427 SOHC Ford-powered Top Fuel car and he wasted no time taking the car to some SoCal Top Fuel

A young and very intense "Snake" in 1967.

Photographer Craig Cutler went on tour with me in 1973 for 36 hours of driving from Los Angeles to Columbus, Ohio.

victories. Mike Snively, driving the Don Long *Hawaiian* No. 3 took Top Fuel honors at the Hot Rod Magazine Championships at Riverside, California, with a 7.07 at 221.66 mph.

It was a good year for Top Fuel shows in California and around the United States. Of course, of the top runners, Don Garlits oddly was not having a good year. He failed to qualify at the NHRA Winternationals and the NHRA Springnationals. The buzz around the Top Fuel pits was, "Has Garlits lost it?" Other Top Fuel teams were breaking into the 6-second zone and Garlits had not. He and his crew vowed not to shave until he broke into the 6s. He again proved he was the King of Dragsters at the 1967 NHRA Nationals because not only did he win Top Fuel Eliminator but he ran a 6.77 at 220.58 mph with his *Swamp Rat XI*, which-sported a 426 Dodge Hemi powered by 100 percent nitro, as well as a slipping clutch, and new M&H tires. He won his victory before a crowd of 20,000 over James Warren, who ran a

It may be old and outdated, but it is a work of automotive art.

Joining the gang of retro Top Fuel cars that began being restored and replicated in the 1990s was the late Gene Mooneyham, who debuted a great replica of his early 1960s Top Fuel car. The 55 dragster was the highly successful ex-Art Chrisman Hustler*. Mooneyham restored it and it was sold to Larry Smith in Texas to be put on display.*

The grand master of drag racing, Mr. NHRA himself, Wally Parks in 1992 was talking with racers and fans at the First Hot Rod Reunion in Bakersfield, California.

Land-speed guru Art Arfons came back to the quarter-mile briefly in the early 1970s when he ran a turbine-jet¬powered dragster at IHRA events. Of course, this wasn't Arfons' first venture into quarter-mile racing. In 1958, his rear-engine Allison-powered dragster set top speed at the Okie City NHRA nationals. The turbine car was around only a short time and didn't set the world on fire.

6.95 at 222.71 mph. The crowd came over the fence as Garlits shaved his beard off right on the Indianapolis Raceway starting line. Everyone knew Big Daddy was back!

Bakersfield-base, Warren, Coburn, and Miller got off to a great start in 1968 when driver James Warren went through a stellar field of Top Fuel cars for a victory in Top Fuel Eliminator at the 1968 NHRA Winternationals. The AHRA Winternationals was claimed by Don Prudhomme with the 427 SOHC Ford Top Fuel car of Baney and Prudhomme. Prudhomme put away the powerhouse team of the Ramchargers in the final. SoCal transplant Ron Rivero drove the former Frantic Four, now K&G Associates, Top Fuel Car to a big win at the Bakersfield March Meet. Jim Fox tuned the full-bodied, 426 Hemi-powered, Pennsylvania-based Top Fuel car. Don Garlits continued his comeback with victories at the NHRA Springnationals and made it two years straight at the 1968 NHRA Nationals. Steve Carbone, driving the Crietz and Donovan Oklahoma-based Top Fuel car, was Garlits' first victim in the final. Speaking of Oklahoma, Bennie Osborn broke into the big-time Top Fuel ranks with his win at the 1968 NHRA World Finals and proved a Southwesterner could race with anyone in Top Fuel.

The racing surface at the 1969 NHRA Nationals at Indy had a lot a bite for Roger Clayton's Top Fuel car. The car went for the moon on two different runs; check out the weight bars clamped on the front axle.

Garlits hit yet another milestone in 1968 when he ran the first 240 mph pass for a Top Fuel

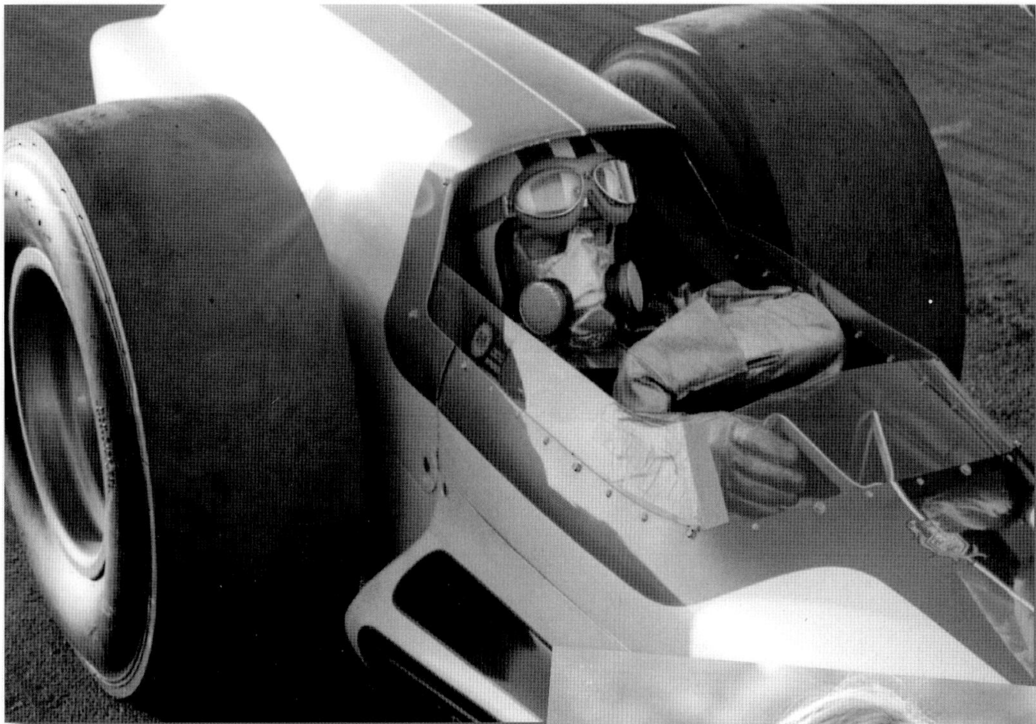

Fling Traylor's USA-1 *turbine made lots of fire and smoke for a great show at night. It wasn't a jet car, but it wasn't quite a Top Fuel car, either. Veteran Top Fuel driver Hank Westmoreland made a few laps and then George Hutchinson took over the driving.*

dragster at Alton, Illinois. Back in Southern California, Doug Kruse, a master aluminum craftsman, became a Top Fuel race promoter and put together a great series of PDA (Professional Dragster Association) races for the Top Fuel racers. Lions, Orange County International Raceway, Carlsbad, Fremont, and Sacramento all hosted 64-car Top Fuel shows under the PDA banner. If you were a dragster fan, PDA races were the place you had to be in 1967-68. Garlits, Jerry Ruth, Vic Brown, James Warren, and Steve Carbone all got nice fat checks from Kruse for their PDA victories.

The 1969 NHRA Winternationals opened up the 1969 drag racing season and Garlits went home a non-qualifier while John "Zookeeper" Mulligan drove the *Fighting Irish* Top Fuel car of Beebe and Mulligan to a Top Fuel Eliminator victory over Don Prudhomme and his new *Wynn's Winder* Top Fuel car. The March Meet at Bakersfield was also won by a SoCal favorite, Jim Dunn in the Dunn and Reath *Rainbow* Top Fuel car. Dunn put away Dave Babler and his California Woody Top Fuel car. Don't feel sorry for Don Garlits—he won the grand opener at NorCal's new Sears Point Raceway, the AHRA World Finals, and Florida's Turkey Trot Drags Top Fuel Eliminator, all with his *Swamp Rat XIII*.

Prudhomme came to the 1969 NHRA Nationals ready to race with his 426 Hemi-powered, full-body *Wynn's Winder* Don Long dragster and took home the Top Fuel Eliminator prize. But all was not happy at the race. John

This full-bodied Top Fuel car remains unknown to me, but it sure was a great example of a retro front-engine fueler. It was at Bakersfield for the Second Annual California Hot Rod Reunion.

At the Second Annual California Hot Rod Reunion in Bakersfield, two beautifully restored front-engine Top Fuel cars were displayed: Don "The Snake" Prudhomme's **Wynn's Winder** ***and Don Garlits'*** **Swamp Rat*****/Smothers Brothers Top Fuel car.***

"Zookeeper" Mulligan , low ET holder at 6.43, was seriously injured in a fire, explosion, and crash during the first round. Once again Top Fuel racers were made aware of how dangerous slingshot-style race cars were. Mulligan died from his injuries two weeks later and the *Fighting Irish* of Beebe and Mulligan was no more.

The strangest Top Fuel win in 1969 was by Larry Dixon driving the Howard Cam *Rattler* at the Hot Rod Magazine Race at Riverside, California. Dixon raced Steve Carbone driving the Crietz and Donovan Top Fuel car. Both had slipper clutches, both left hard in the final, and Dixon was smoking the tires as Carbone's car hooked up and away he went past Dixon. With a giant lead, Carbone's right arm was standing straight out of the cockpit and his gloved hand flashed a V for victory. Meanwhile Dixon was backpedaling and driving the *Rattler* all over his lane. Just past the eighth-mile Carbone's car lost power from a broken driveline, and with tires blazing, Larry Dixon went by a powerless Steve Carbone for the win. However, 1969 closed with Steve Carbone taking the Crietz and Donovan car to a Top Fuel win at the NHRA World Finals.

In the new decade of the 1970s, Top Fuel dragsters had to share the nitro glory with the now official NHRA Funny Car class. The SoCal tracks would alternate between Top Fuel and Funny Car shows that would either combine Top Fuel with Fuel Altereds or Funny Cars with Fuel Altereds. Larry Dixon, driving his and his wife's family's Dixon Top Fuel car, stopped Tony Nancy's *Wynn's Sizzler* for Top Fuel honors at the 1970 NHRA Winternationals, and this would end up being the last time that a slingshot dragster would win the NHRA Winternationals. At the March Meet in Bakersfield Tony Nancy redeemed his final round loss to Dixon when he drove his

426 Hemi-powered *Wynn's Sizzler* past "Hand Grenade" Harry Hibler in John Smyser's 392-powered digger to win the March Meet. It would be Nancy's only national event win.

Over the years the death and injuries were starting to mount up and drivers began to look for a better, faster, safer race car. However, March 8, 1970, was the beginning of the end for the slingshot-style Top Fuel dragster as this day was marked with the transmission explosion and injury to Don Garlits at Lions Drag Strip in Long Beach, California. Garlits was in *Swamp Rat XIII* and had a transmission failure that put him in the hospital with part of his right foot cut off. While recovering from his injuries, he decided that a Top Fuel driver must be put out of harm's way, so he and Connie Swingle hit the drawing board to come up with a rear-engine dragster. Sure, there had been rear-engine dragsters before, but making them work was another story.

A new race on the NHRA National trail was the Gatornationals at Gainesville Raceway in Gainesville, Florida. Dave Chenevert drove Robert Anderson's Louisiana Top Fuel car to a victory

If it runs on nitro, Gerry Steiner has driven it. Steiner even drove the redone 554 Coupe. In the early 1990s Steiner donned his black mask and slid behind the wheel of a 392-powered retro Top Fuel car for a trip down memory lane.

One fixture at the Lions Drag Strip and frequent visitor to the winners circle was the SoCal Top Fuel team of Yabsley and Mitchell. The Red Mountain Boys *Top Fuel car was ably driven by little Johnny Mitchell and they were a fan favorite at Lions. Always the underdog, the team always seemed to be in the money rounds by the end of the evening.*

Pete Kaiser's blown Chevy was one of the first retro front-engine Top Fuel cars. Kaiser was a lot of fun to photograph and watch in the early 1990s.

over Illinois-based Jim "Yankee Pack Rat" Paoli. Both cars were of show-and-go quality, full bodied and sporting 426 Hemis. After Gainesville, the racers headed to the Lone Star State of Texas for the NHRA Springnationals at the Dallas International Raceway. Local boy Bob Gibson and the Carroll Brothers, a.k.a. the Texas Whips, made good and sent all the Top Fuel invaders back home. With a Texas flag flying from the T-bar chassis rollbar, Gibson piloted the 426 Hemi-powered dragster past SoCal's Jim Nicoll and his sleek *Der Wienerschnitzel* Top Fuel car for Top Fuel glory. With the better tires by M&H and Goodyear and of course slipper clutches, the smoke-filled runs by the slingshot dragster were going the way of the dodo bird.

The Summernationals was all Ford with "Sneaky" Pete Robinson's *Tinker Toy* Top Fuel car with ts 427 SOHC Ford engine defeating Jim Nicoll for top honors in Top Fuel. The NHRA Nationals would once again prove how dangerous slingshot dragsters could be. The final round pitted old buddies Don Prudhomme and Jim Nicoll against each other and when the Christmas tree flashed green in the final, Prudhomme and Nicoll left side by side. It

almost looked like a dead heat at the 1,320 mark when all hell broke loose in Nicoll's yellow Top Fuel car. Parts, pieces, smoke, and fire as Nicoll and his rollbar and rear wheels flew over the guardrail bouncing wildly on the grass between the guardrail and spectator fence. Meanwhile on the race track Prudhomme was slowing the *Wynn's Winder* down from its victory run. Prudhomme looked over to see where his buddy Nicoll was, and Nicoll's car or what was left of it slid by still in its own lane. Nothing from the back of Nicoll's 392 Hemi was there, just the front wheels, body, and engine, and Prudhomme was sure that his good friend was dead. Unlike with John Mulligan, this would be a happy ending, Nicoll walked away without a scratch despite a monster clutch explosion and he earned a new nickname, "Superman."

Wrapping up the 1970 season was Ronnie Martin driving Robert Anderson's Top Fuel car and defeating "Kansas" John Wiebe and his 417 Donovan-powered Top Fuel car at the NHRA World Finals. Towards the end of 1970, the NHRA added another national event, the Supernationals at Ontario, California. Local Top Fuel hero Rick Ramsey drove the beautiful Keeling and Clayton *California Charger* with a 392 Hemi to win the first ever Top Fuel title at the Supernationals.

In December of 1970, Garlits and Swingle began to test a new rear-engine Top Fuel car at home in Florida. Garlits still wasn't sure about the new design and had a brand new slingshot ready to go if tests weren't positive. The 1971 drag-racing season opened with the NHRA Winternationals and the buzz in the pits was about Garlits' new rear-engine Top Fuel car. This wasn't his new car's first appearance on the West Coast as it had run in early January at OCIR and was at the Lions AHRA Grand Opener three weeks before Pomona. The car ran well but was nothing to rave about. Whatever bugs his new car had, Garlits and crew found them and they marched through a stellar field of slingshot-style Top Fuel cars and finished off with a victory over Kenny Safford in Larry Bowers' "new" slingshot-style Top Fuel car. The Top Fuel community was stunned. Was this a fluke? All of 1971 was a battle between the two dragster designs. At the March Meet in Bakersfield, California, Garlits raced his old nemesis Don Prudhomme in his *Wynn's*

CarTech author Bob McClurg, shown at the 1971 NHRA Summernationals, works hard to provide the best in photography and writing.

Winder Top Fuel car. It was a classic battle of the new and the old, neck and neck they raced down the Bakersfield 1,320 and at the eighth-mile Garlits' arm rose out of the car and a V for victory was seen on his hand—and just at that moment there was a huge blower explosion from Don Prudhomme's old-style car. While Prudhomme fought for control of his car, Garlits streaked through the finish line. Prudhomme eventually stopped but was minus an injector and blower while Garlits went on to collect Top Fuel honors defeating Rick Ramsey in the final.

By the end of 1971, the writing was on the wall. Garlits ran 16 of the major drag races that year and won 11. He was also runner-up in another four and lost in the second round in one. Garlits also took home AHRA World Champion in Top Fuel that year as well. It was time for the Top Fuel class to wake up to a new era of drag racing because the slingshot was now obsolete.

The "other" Jack Williams campaigned this beautiful Syndicate Scuderia AA/D from Vancouver, British Columbia. This very cool car ran in the mid 1960s.

This Jim Davis chassis was one of the great buys in 1992 and was a great start to a retro Junior Fueler although it would have to be updated to run current Hot Rod Reunion rules. A Jim Davis chassis would be a nice score for a would-be nostalgia racer as Davis passed away and there is no possibility of getting another one.

Before the Chrysler Hemi, the Ardun was the "hemi" of choice. The "Grey Fox," Jack Ewell, made his Ardun-powered dragster thunder in the 1950s.

A member of the Inland Roadsters of SoCal stops and remembers the good ol' days when Ardun dragsters were the kings of the quarter-mile.

Mickey Thompson's Panorama City Special *is still a beautiful piece of four-wheeled art. All Top Fuel racers learned a great deal from Thompson; he was a true innovator.*

One Sunday afternoon in 1965, a beautiful new Top Fuel car appeared at Fremont Drag Strip. Raced by the Young brothers, this blue beauty was only at a handful of races before it disappeared, never to be seen again.

In June of 1965, the Team of Albrich, McCulloch, and Krieger, with Ed McCulloch driving, beat "Sneaky Pete" Robinson for the Drag News No. 1 spot in Top Fuel. McCulloch beat Robinson two straight to claim the title of No. 1 Top Fueler with a best of 7.5 at 206.42 mph. The Northwind ruled!

Getting the action on film was something the late Leslie Lovett could do well. NHRA's premier photographer brought out the best of any NHRA event he covered. He was one of the guys I learned a lot from in my early days of shooting racecars. His sudden passing in 1996 was a true loss for drag racing.

Engines, engines, and more engines the theme in Top Gas eliminator in the late 1960s and early '70s. Jack Jones, Gordon Collet, and Motes & Williams were prime examples of Top Gas gone wild. Top Gas was getting good when NHRA killed it in the early 1970s.

A great find in the swap meet area at Bakersfield in 1992 was the original Speed Sport *roadster, a real piece of drag racing history. The* Speed Sport *roadster, an early rear-engine racer, was campaigned by Lyle Fisher, Red Breth and Don Maynard and was one of the most popular cars of its era. Between 1956 and the early 1960s the red roadster ran 160.14 mph at the 1957 ATAA World Series and ran a 169 mph at Tucson, Arizona. Other records include an 8.03 at 175mph in March of 1960, a 180.72 mph backed up by a 179.28 mph on October 2, 1960, and the Drag News 1320 Standard Modified Fuel Roadster speed mark of 183.67 mph eight days later in Alton, Illinois. All this was from a blown 354 Chrysler Hemi in the 1,500-lb roadster. The original car made its final appearance at the 1964 March Meet in Bakersfield.*

The freshly restored Speed Sport *roadster was seen briefly in 1974 at the AHRA Winternationals when Greth and Fisher restored it. The car was found again at the Hot Rod Reunion in the early 1990s and then it was restored again to rest in the NHRA Motorsports Museum.*

One of the last Drag Master slingshot dragsters to run in Northern California was raced by Barnard, Lapham, and Davis. The Drag Master chassis was one of the most popular in the early 1960s. Jim Nelson and Dode Martin were the country's most prolific dragster chassis builders from the late 1950s and into the 1960s. Their customers included Roland Leong, Danny Ongais, Pete Robinson, and Mickey Thompson. When Drag Master unveiled its Streetster Hot Rod kit in the mid 1960s, more than 50,000 letters of inquiry poured in. The company still exists today as a specialty automotive shop.

NORCAL, HOME SWEET HOME

Wally Allred's Top Fuel car is a great example of the state-of-the-art in the early 1990s. It was typical in that it was full bodied with a 392 Chrysler Hemi engine but it differed from the original slingshots because it had 1990s innovations built into it. The chassis was built safer with thicker tubing, it had a double roll bar with a better clutch can, safety straps on the blower and any other safety measures that had to be implemented at that time.

With Ted Gotelli having to regroup following partner Denny Milani's death, the battle for Top Fuel king of NorCal began. Builder and driver Jim Davis seemed to be the best choice to rule the Top Fuel roost. The up-and-coming John Batto was already fading from Top Fuel, but Dennis Baca, a customer and friend of Davis, was making waves in the class. Also flexing his muscles was a young kid from Oklahoma, Gary Ormsby. Ormsby started driving the *Vagabond* Top Fuel car in 1965 when he was 24 and was now in the seat of the Howard Cameron-tuned *Cheetah*.

At least four injected fuel diggers were ready to give the blown fuel boys fits. The injected cars were lighter and left the starting line faster, but the blown cars had more HP and could make up ground on the top end. Leading the list was the Woody-chassis *Vic Hubbard Speed Shop Special* driven by Denny Forsburg. The Chevy fueler cranked out low 8s, and driver Forsburg could leave on the best of them. Jim Brandy's *Giant Killer* was a leaver, too. The Sacramento-based Brandy enjoyed putting the blown fuel Chryslers on the trailer. George Santos' *Goldfinger* Chevy with driver Jim Perry had quite a few notches on its belt as well. Running out of S&S Automotive in Hayward, Santos was longtime NorCal racer and sponsor of drag cars. The Brothers Galli had the only Oldsmobile-powered Top Fuel car, but not for long. A nitro-gulping Chrysler soon replaced the injected Oldsmobile.

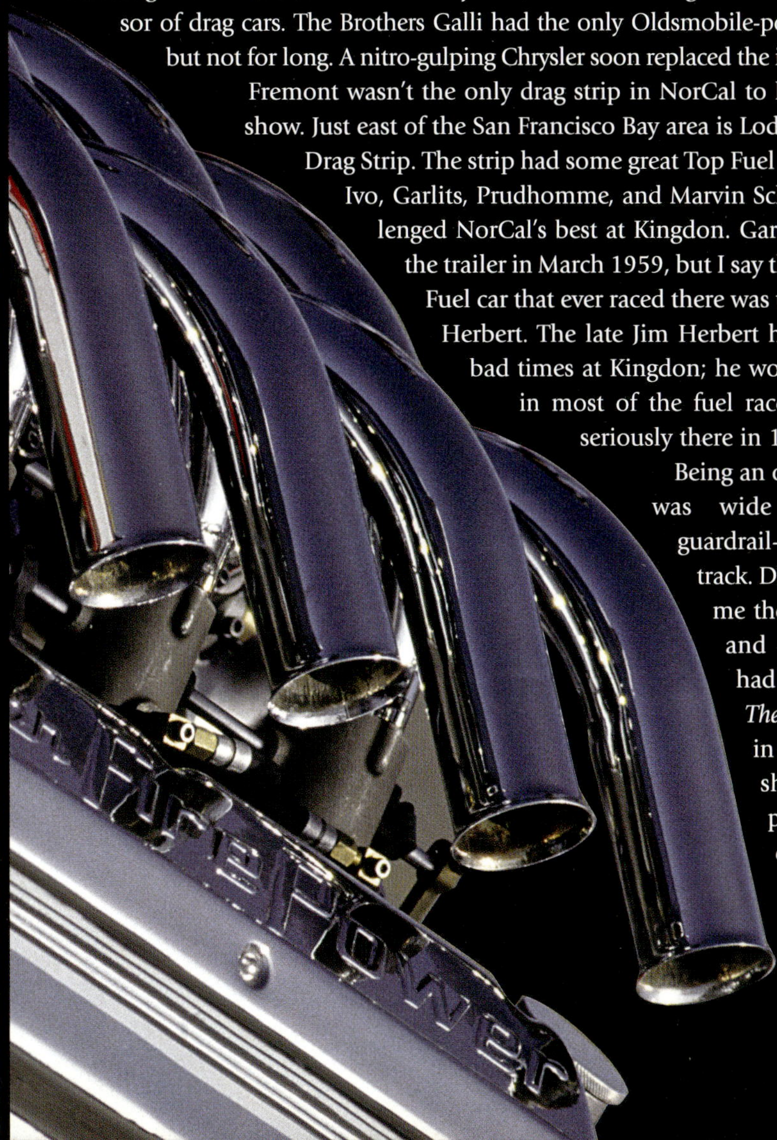

Fremont wasn't the only drag strip in NorCal to hold a big Top Fuel show. Just east of the San Francisco Bay area is Lodi, home of Kingdon Drag Strip. The strip had some great Top Fuel shows in the 1960s. Ivo, Garlits, Prudhomme, and Marvin Schwartz had all challenged NorCal's best at Kingdon. Garlits put them all on the trailer in March 1959, but I say the best-running Top Fuel car that ever raced there was the Pitts & Bishop & Herbert. The late Jim Herbert had good times and bad times at Kingdon; he won or was runner-up in most of the fuel races, but was burned seriously there in 1968.

Being an old airport, Kingdon was wide open with no guardrail—my kind of race track. Don Cook tried to kill me there in his *Southwind* and Marvin Schwartz had his steering fall off *The Anaconda* with me in the shutdown area shooting parachute photos. Schwartz came whizzing by me with the front

My very first sighting of this Top Fueler was in 1964 when Tom Pruffer owned it and Gerry Steiner drove it. The Power King *was a very cool fueler. This full-bodied beauty was very much a show-and-go racer.*

The Whiz Kids Top Fuel car of Bumgarner, Trefethen, and Pearson had a jet fighter pilot at the controls. Walt Bumgarner took time out from flying Uncle Sam's fighter jets in Vietnam to drive the car.

wheels flapping around and the steering dragging on the ground. One other thing about Lodi —it was the coldest *and* the hottest place I have ever photographed racecars. It was 36 degrees on January 1, 1968, and the same year in summer it was 122 degrees.

A rather short but scenic drag strip came from another airport turned race track in Half Moon Bay, California. The track was just south of San Francisco and a stone's throw from the Pacific Ocean. It was an old place that needed a serious renovation. The first time I brought Los Angeles-based photographer Tim Marshall to Half Moon Bay, he couldn't believe what he saw. He was amused by the field mice running all over the old wooden grandstands and the uncut weeds surrounding the track. To be honest, I thought anyone who ran a Top Fuel car there was nuts because when you go fast, you need lots of parachute and brakes, and this track was very short. Bob Haines, driving for Masters

The first Woody Gilmore car in Northern California was raced by John "The Sultan" Batto. The official team lettered on the cowl was Batto, Valente, and Bings. The team raced out of Santa Rosa, California, and in 1965 Batto blasted out an unheard-of 7.69 at 213 mph. Prior to that, the team's claim to fame was crewman "Sonoma Fats," who ran over the Christmas tree at the 1964 NHRA WCS points race. Fats was angry because of all the red lights given out, so he jumped into the BVB push car and, crunch—no more Christmas tree. I saw the whole thing. He received an ovation from the crowd.

and Richter, had problems stopping there during the early 1960s. The Top Fuel car ended up at the end of the track, upside-down in ditch full of water. Haines almost drowned. Of course, Denny Milani had a chute problem while driving the Gotelli/Milani Top Fuel car in August 1965 that cost him his life at the end of Half Moon Bay. Still, some of the finest in Top Fuel continued to race there. Garlits and Prudhomme had huge wins, and even Gotelli returned with his new Woody Gilmore-chassis car and driver Kenny Safford a year after Milani's crash.

Just outside California's capital city of Sacramento is Sacramento Raceway. It was built right next to the Air Force base and opened in the mid-1960s. Milani and his Ogden-chassis Top Fuel

In 1965, John Batto introduced the first Woody into NorCal Top Fuel racing. The **Sultan of Santa Rosa** *quickly became the car to beat.*

car won top honors at the track's opening event on May 15, 1965, beating Bobby Tapia in the Stellings and Tapia entry in the final. Both cars ran 392 Chrysler Hemis.

For some reason, I was never comfortable photographing race cars at Sacto, but some amusing racetrack stories have come out of there. The owner/manager's home was on the premises. He also had his own plane there. One evening after a rather large race, a dispute arose between a group of racers and the owner. The argument got very heated and both parties left the table angry. The racers had a plan to get back at him, and in the middle of the night, the group of disgruntled racers snuck back into the track and pushed the small private plane out of its open hangar toward the open backyard of the home, and with one big shove, the plane was deposited into the family's swimming pool. As the plane slowly sank to the depths of the pool, the racers headed for their parked getaway trucks on the outside of the race track. Revenge was sweet.

Other smaller tracks in NorCal were also interesting. Just south of Sacramento is the small town of Vacaville, just off of Interstate 80 in the middle of onion fields. It had a small drag strip with a big heart that was very popular with the slower-class racers, but Top Fuel and Top Gas didn't fare well with the short shutoff area. Then there was the track in Redding, an old airstrip with a makeshift scaffold for a tower. Salinas sported another old airport-turned-racetrack that had cows for spectators on the right side. They periodically stopped racing so that the cars could return to the pit area from the end of the track. It had to have had one of the most eclectic tower/announcer booths I have ever seen—an old milk truck.

Heading south to Central California, Fresno Dragway was a long, wide track that had some great Top Fuel shows in the late 1960s. Then of course there was Famoso Raceway in Bakersfield, another old airstrip that was wide and long and a favorite battleground for the Top Fuel dragsters from NorCal and SoCal and, oh, that guy from Florida, Don Garlits.

— FOLLOWING THE SWEET SMELL OF NITRO SOUTH —

By 1963 I was a drag-race pit rat. Fremont Drag Strip was my home away from home, and at the end of 1965 I could recite stats on every Top Fuel dragster in NorCal, and I had photos of every one. Many of the photos were being published by the local eight-page *Drag Gazette,* which was distributed at local drag strips by the editor/publisher, Al Caldwell. I was also targeting *Drag News* with my photos, but hadn't yet been successful.

Then Jon Asher, the editor of a new magazine in the Midwest called *All American Drags,* got me a pass to the largest Top Fuel gathering on the West Coast—the Mickey Thompson 200-mph Club meet at Lions Drag Strip. Lions was about 400 miles from my home in Fremont and I had been to only two other drag strips outside of NorCal. I had ventured to the *Hot Rod* Magazine Meet in Riverside with Rich and Les Welch, where I had no press pass, but I snuck out to shoot pictures anyway. Then we went to Fontana,

Lions Drag Strip during the late 1960s, with the **Howard Cams Special** *in "the zone," as Lion's Drag Strip's tunnel effect could put most drag racers into deep thought or for that matter deep fear.*

A team that brought fear to SoCal Top Fuel teams in the mid 1960s was the "Jungle Four," of Mooneyham, Jackson, Ferguson, and Faust. These guys terrorized SoCal drag strips from Fontana to Lions, taking more than their share of Top Fuel Eliminator victories. "Jungle Larry" Faust was the very capable driver of the blue beauty.

FOLLOWING THE SWEET SMELL OF NITRO SOUTH *continued*

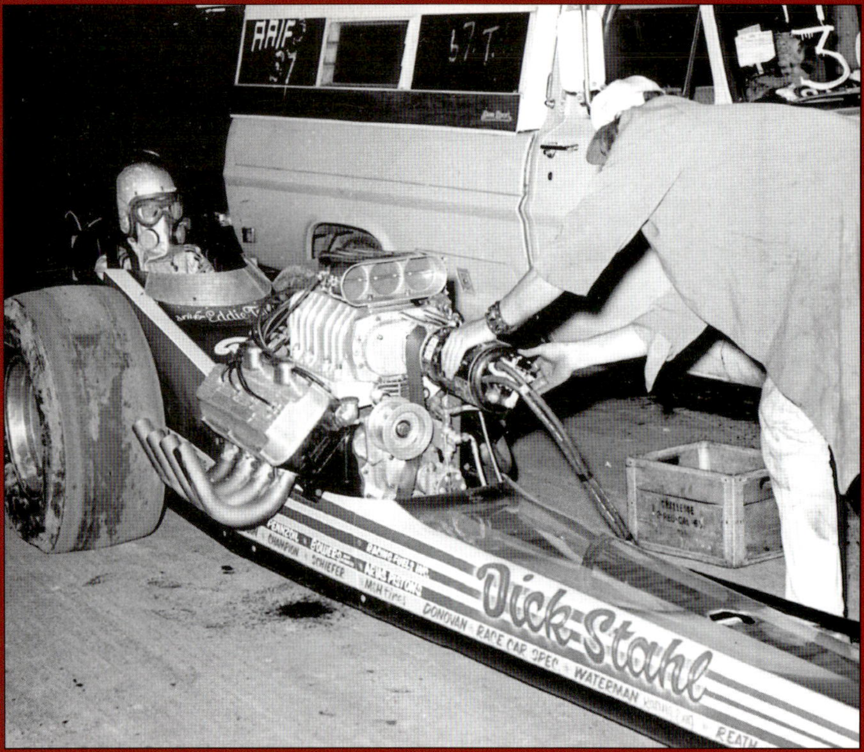

The colossal human being preparing to start his Top Fuel car is the late Dick Stahl. He was a 7-ft giant in SoCal Top Fuel racing in the mid to late 1960s. Heir to a local fast-food fortune, biker, and a man who had been married eight or nine times, Stahl was the easygoing giant in Top Fuel. An owner of a blown fuel-burning roadster, Stahl wanted to go faster, so he entered the Top Fuel class with driver John Wilson. Between them they had one of the strangest Top Fuel runs ever at Lions Drag Strip. On a Saturday night, Stahl and Wilson went up against Shubert and Herbert. The pair staged, the green light flashed, and the race was on into the finish-line lights when a large explosion engulfed the Stahl and Wilson car. Both cars stopped safely, only the Stahl and Wilson car had been modified a bit: no engine, no bellhousing, no fuel pump, no firewall, and it only had three wheels. Wilson walked away fine in his newly lightweight three-wheeled fueler.

where I watched Bill Martin's *400 Junior* with Gary Cassidy driving win Top Fuel on Saturday night. The upcoming trip to Long Beach was going to be very different. I would be shooting for a magazine!

I loaded up and headed to "The Beach," along with my photo buddy Ted Stewart and his 354 Hemi Desoto. It was a six-hour drive and we couldn't leave until after Stewart finished work on Friday. About midnight, tired yet excited, we entered the Los Angeles city limits chatting about film, cars, and camera angles. Suddenly, grinding and strange noises overtook the mighty Desoto. Stewart nursed the wounded car off the freeway until it finally died. It was after midnight, we were almost 400 miles from home with a broken car, and guess what? We were in the Watts area of Los Angeles, home of the 1965 Watts riots. It was a bad place to be wandering around with our suitcases and camera equipment in the middle of the night.

We came across a guy trying to get his car started, and we agreed to give him a push if he would give us a ride to Long Beach. So we push this guy's car, it starts, and he waves goodbye, leaving us standing there in the middle of the street looking at each other. We gathered up our goodies and kept walking. We finally caught a bus, and in about 30 minutes we were in downtown Long Beach. By then it was about 3

A.M. and we were once again wandering the streets, this time in Long Beach. Even with pooling our money, we had very little for a hotel. We found a place in downtown Long Beach for $5 per night, a flop house. There were lots of bums and drunks to step over to get to our "room," and the sweet smell of cheap wine and stale urine hung in the air. It was 4 A.M. and we needed a couple of hours rest. Stewart went for the bed and I got the floor, yes, the floor of a flop-house hotel room. We woke about four hours later, and I checked myself for major bug bites, washed my face, and headed for the nearest bus stop.

The bus dropped us off right in front of Lions Drag Strip at 9:30 A.M.. We finally made it to Long Beach. During the daylight hours, Lions looked like a collection of poles, wires, grandstands, and asphalt. But at night, it became the Roman Coliseum. The Romans (fans) were in the massive grandstands overshadowing the quarter-mile asphalt battleground. The gladiators were in their helmets and fire suits, strapped into long high-HP missiles. We witnessed more than 15 hours of Top Fuel, Top Gas, and Junior Fuel cars, and at the end of the evening, with a thick smell of nitro and rubber in the air, Don "the Beachcomber" Johnson with driver Floyd Lippencott Jr. had taken home Top Fuel honors.

There was nothing anywhere in drag racing like Top Fuel qualifying at dusk at Lions Drag Strip during the late 1960s. Walt Stevens drove the Trip Top Fuel car. **The Trip** *was one of the many rides for Walt Stevens who was also the driver for the infamous* **Odd Couple** *Top Gas Slingshot. The* **Odd Couple** *got its name from its engine combo, a blown Chevy and blown Chrysler Hemi in the same car at the same time.*

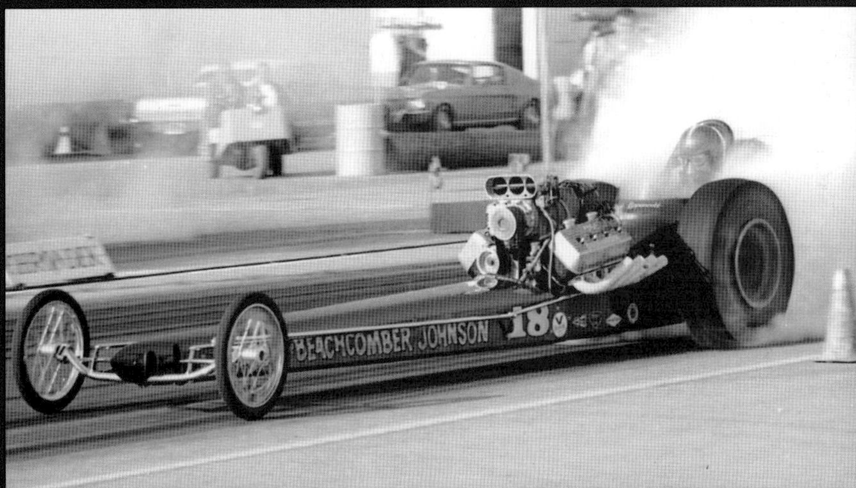

Don "The Beachcomber" Johnson showed up at the 1967 NHRA Winternationals with a very nice-looking racer; Johnson had spent some of the loot he won the latter part of 1966 at the Mickey Thompson 200 mph Club race. The mythical Floyd Lippencott Jr. drove for Johnson.

FOLLOWING THE SWEET SMELL OF NITRO SOUTH *continued*

Look closely at the front of the **Sugar Cane** *Top Fuel car and you will see the crew had mounted a movie camera to record their assault on the Lions quarter-mile. The problem was that the camera shook open and the Top Fuel car ran into the 8mm film coming out of the camera. The driver had to shut off and a nice mess of film had to be cleaned up.*

To be honest, I don't recall how Stewart and I got home, but we did, and it was one great first time at Lions Drag Strip for both of us. Back home at Fremont Drag Strip, lights had been installed and drag racing moved to Saturday nights in NorCal. Following the death of Denny Milani in August 1965, the Top Fuel class struggled to recover from the loss. Night racing helped give the NorCal drag fan a different view of Top Fuel.

The German Army, Fremont, California, branch? Naw, just photographer Ted Stewart. He put up with a wide-eyed kid photographer (me) in the mid 1960s. We always had adventures when we traveled together. We were mistakenly detained for armed robbery when we ended up in Watts at 2 A.M. with a blown-up car. At a $5-a-night flophouse in Long Beach, I got the floor and Ted got the bed. We had crazy times, but always got our photos taken and published.

The Junior Fuel digger staged, the light flashed green, the driver hit the throttle, and drove straight into the Lions Drag Strip guardrail. No harm, just a very bent racecar and a driver who wants to remain unknown.

A mainstay at Fremont Drag Strip in the early to mid 1960s was Terry Ervin and his Junkyard Special B/FD. It looked like a moving trash pile, but it won races.

One of my early favorites was the tire-churning '34 Ford-bodied Charlie's Business Coupe, *a real crowd pleaser in the mid 1960s. Lots of smoke and wheels up in the air runs were a couple of Charlie's trademark. It didn't win a lot but the fans loved it.*

Instead of guessing who this race team is at Lions Drag Strip in 1967, let's just say it is another Top Fuel car testing the bite at the famed Long Beach, California, track.

California versus Georgia at Lions in 1967: Dave Beebe and "Sneaky Pete" Robinson squared off at "The Beach." Beebe won the close, tire-smoking race.

NorCal had all kinds of cars running in Top Fuel. This blown, fuel-burning Ford is the Flanders Ford digger.

Two NorCal heavyweights in Top Fuel square off for Top Fuel Eliminator. Jim Davis (near side) holeshots John Batto and smokes 'em to the finish line for the win in 1966.

It's Chevy versus Chrysler in Top Fuel with John Batto (far side) and Frank Bradley. Batto won the very close race.

Northern California had one haulin' blown fuel Chevy Top Fuel car in the Broadrick, Bradley, and Swansboro nitro burner. It battled the Hemis every weekend. This was Frank Bradley's first ride in Top Fuel, and he did well.

Another NorCal smoke machine was Gerry Card's AA/Fuel coupe. This silver fuel coupe/dragster was Hemi powered and ran against NorCal Top Fuelers in the mid-to-late 1960s.

Now you know Prudhomme's secret for winning, he personally ran Pennzoil in him and his racecar.

A fixture in the Top Fuel class in Northern California was "The Texas Lizard," Jim Davis. Based in Walnut Creek, California, Davis was an innovator, building race cars, and driving them, too. Davis came up with the first starter system for blown fuel engines. A bad crash in his first years of racing broke his neck. Most people would have quit, but not Davis. He once more took the controls of his Texas Lizard *Top Fuel car. The key to Davis' remarkable comeback was his wife, Carol.*

Northern California's answer to the Logghe Brothers slot racer was George Santos' Gold Finger *A/FD. Jim Perry drove the sleek Chevy-powered A/FD, beating many a blown fuel-burning Hemi and Chevy digger.*

Bob Nestor and Don Durbin made the jump from a B/Dragster to a thundering front-engine, nitro-burning Top Fuel car. The Newark, California, duo raced mainly in California. While they never became nationally famous, they had a lot of fun.

The Fulmer Brothers had a better idea, a rear-engine A/FD. Their car didn't set the world on fire in the mid 1960s, but it was a sign of things to come in nitro dragster racing. It would have run better if the fuel leak had been fixed.

Ed Gonzales' The Mixer *was a show-and-go A/FD powered by a Chevy and running in Top Fuel Eliminator, which mixed Top Fuel, Top Gas, and A/FD racecars. Terry Ervin and Larry Johnson drove the Hayward, California-based race car.*

"Little guy" Top Fuel racer Jimmy Haire came out and competed with the best NorCal had in the mid 1960s. Haire did okay with what he had to work with.

This nice-looking Top Fuel car came from Sacramento to race at Fremont in the mid-1960s. The Wulf and Maher car wasn't at Fremont often as it was a fairly long tow from Sacto. I believe Greg Maher drove at first and then "Shorty" Leventon took over for him at some point.

Rudy Babino's **Velvet Hammer** *goes "poof!" on the starting line at Fremont. Early slingshot Top Fuel cars were simple but often had gremlins that would driver the owners and drivers crazy. A simple engine backfire could put the car on the trailer or in the pits for repairs.*

Fremont flagman Bob Gewertz gives the go sign to Dick Bauer in the Bauer, Blessing, and Swansboro Chevy Top Fuel car circa 1965. This Chevy dragster had success when Frank Bradley took over the driving chores. The Chevy-powered dragster could race head to head with the 392 Chrysler Hemis of the day, much to the delight of the Chevy fans in Northern California.

A true class act in the NorCal Top Gas class was the team of Goodknight, Keith, and Williamson. With Bob Keith (shown) at the controls, this San Jose, California-based team gave the Hemi-powered cars fits. These guys were great because they took care of me every time I got their photos published in any magazine or paper. Since I was a struggling kid photographer, the extra $20 was very helpful.

The AA/MFR class was exclusive to Southern California. Take a Top Fuel dragster with a small Hemi engine and put a modified '23 T fiberglass body on it, and away you go. I only saw this class run at Lions, but occasonally these "roadsters" competed in Top Fuel at Irwindale or Bakersfield. When Lions died, so did this class of "Nitro Roadsters."

DRAG WORLD

VOL. 4, NO. 21 — FRIDAY, AUGUST 30, 1968

FEATURE INDEX

DON COOK WINS AHRA NATIONALS

Easygoing and always with a big grin on his face — that was Don Cook. I first met Cook in 1965 when he was teamed with Dwight Bale in the Cook and Bale Top Fuel car. He soon entered the NorCal Top Fuel wars solo. His Jim Davis chassis racer soon was making waves and Cook became "Mr. 223," a nickname he earned with his mph clockings. Within a couple of years, Cook was an AHRA Top Fuel touring pro. In August 1968, he won the AHRA Nationals with buddy Leroy Goldstein's engine in his racer.

Not only did Bob Gewertz start racecars, he was a partner in one, the Top Gas car of Arigoni, Grigg, Gewertz, and Bernasconi. Gewertz made a few passes in the Hemi-powered digger.

Roy Dunn's blown fuel Chevy was no match for John Batto's blown fuel Hemi during Top Fuel action at Half Moon Bay in 1965. This Hemi-over-Chevy (or anything else) scenario was repeated quite often.

The Galli Brothers started in the A/FD class with a nitro-burning Olds in their digger, but they soon discovered it couldn't compete with the Chevy and Hemi-powered A/FDs, so a Chrysler Hemi adorned the NorCal A/FD, and Rich Galli was driving to some impressive speeds and ETs.

Northern California rivals Jim Davis and the Whiz Kids. It was a cool winter afternoon in 1965 and the smell of nitro filled the Fremont air.

Dennis Baca of Walnut Creek, California, was first paired with Dwight Bale. They ran a Jim Davis-chassis Top Fuel car and took turns driving. That partnership didn't last long and both went on to race their own Top Fuel cars. Baca had great success in the late 1960s and early 1970s.

Half Moon Bay 1965: John Zoucha's Gateway Chevy is making a hard charging run with Frankie Silva at the wheel. This yellow beauty could be found racing throughout NorCal in the mid 1960s.

This is the injected nitro-burning Chevy that no blown Hemi-powered Top Fueler wanted to race. The Vic Hubbard Speed Shop Special *with Denny Forsburg at the wheel was one bad hombre in Top Fuel. This Hayward, California-based entry put many a Hemi Top Fueler on the trailer in the mid 1960s.*

Pure Fury *was what Dave Dozier named his injected Hemi-powered A/FD. This photo is from August 1965 and Dave is running the snorkeled Hemi in the Top Fuel class.*

A personal favorite was "The Flaming Mexican" Frank Pedregon. His Taco Taster *coupe was a crowd pleaser wherever it ran. Here Pedregon is running the coupe in Top Fuel Eliminator, defeating Bob Hall and the* Capitol Speed Shop Special. *Pedregon didn't run much in NorCal so I only saw the coupe a couple times.*

"The Wolverine," the late Denny Milani was hands down the fan favorite in NorCal Top Fuel racing. His fearless driving style awed and thrilled drag fans all over Northern California. When he and his partner Ted Gotelli took delivery of their state-of-the-art Pete Ogden slingshot, big things were in store for the San Francisco duo. The most innovative thing about the Ogden-built car was that the driver's legs went under the rear end not over it. Just when the duo was on top of their game, a chute failure at Half Moon Bay in August of 1965 proved fatal for Milani.

Davis (far side) raced Frank Bradley in another Chrysler versus Chevy battle — just like the ones that occurred almost every weekend at Fremont in 1965 and 1966.

The late Jim Davis was a chassis builder, engine builder, and a tough-as-nails Top Fuel driver. The cars he built raced all over the USA and included Dennis Baca, Dwight Bale, Frank Bradley, the Whiz Kids, Galli Brothers, Ed Roth, etc.

It was December 1966 and "T.V. Tommy" Ivo had just competed at Half Moon Bay—a short drive from San Francisco's North Beach. Bill "I can talk my way into anything" Wishart convinced the Off-Broadway Topless Club they needed to have Ivo's Top Fuel car on display in front of their nightclub, and then talked Ivo's half brother Don LaForce into driving Ivo's rig into San Francisco. Where was Ivo? He was on an airplane heading home to SoCal and didn't know anything about it. Of course, I volunteered to take photos. Just the ride through San Francisco was an adventure. Ivo's rig was not small and Don had his hands full, but we finally arrived and unloaded the dragster. The next thing I knew, a very busty blonde was asking me where I wanted her. After my 18-year-old mouth closed, I instructed her to climb aboard the car. She was Yvonne D'Angers, Miss Topless 1966. We shot our photos, lots of photos, with people coming by to look at the car and model, and many cars drove by honking to let us know they liked what they saw. The manager invited us in for Yvonne's show. I was underage and so was Don, but that was no problem; he escorted us in the back door and there we were front and center; a couple of minutes later, I found out why Yvonne was Miss Topless 1966. After the show, we said a fond goodbye to our new topless friend and loaded up the car—Don headed home to Los Angeles and Wishart and I headed home to Fremont. I still don't know if Ivo ever found out. I guess he knows now.

"Wild Bill' Alexander (far lane) piloted the Hemi-powered Cheetah *to victory over Danny Brown and the Bauer, Blessing, and Swansboro blown Chevy at Fremont circa 1964–65.*

In the mid 1960s Don Tognotti had a beautiful Top Fuel show car called The Bushwacker. *After competing on the show car tour it was sold to Marc Danekas, who promptly put the car on the drag strip. Jack "Mosquito" Martin drove the green-and-yellow beauty.*

This is T.V. Tommy Ivo's last slingshot Top Fueler, shown at the NHRA Nationals in 1971.

Dreams do come true; at least they did for wheel-chair-bound Bill Wishart of Fremont. Wishart's dream of owning a Top Fuel dragster came true in the late 1960s with a sleek Pete Ogden chassis driven by "Wild Bill" Alexander.

Harry and Maxine Lehman ran a hard-charging Top Fuel car on the East Coast during the late 1960s. The Lehmans' American Way *was driven by the late Al Friedman and could be found match racing on the East Coast as well as attending NHRA national events east of the Rockies. Their 20-something driver was a great guy and always said hello with a big smile.*

Coming all the way from Rockford, Illinois, was Ron Colson and the Stiletto *Top Gasser. Here Colson takes on the* AGG&B Special *at Fremont. About one and a half years later in 1968, Colson was at the wheel of the Petersen and Colson Top Fuel car in Bakersfield, California.*

Mike Mahay is caught in the classic flagman's pose as "Wild Bill" Alexander red flags the Cheetah *Top Fuel car. Mike Mahay was a very interesting flag starter at Fremont. He was a pilot who flew cargo planes in the early days of the Vietnam conflict. He also raised piranha at his house in a huge old converted down style freezer. Fremont used flagmen until 1966 because of the expense of the Christmas tree system. Finally a group of drag strip owners got together and purchased the Christmas tree system and it traveled around the NorCal tracks. "Wild Bill" Alexander was a longime veteran Top Fuel driver from Southern California who drove numerous SoCal Top Fuel cars. The most famous being the* Ernie's Camera Shop Special*. At the time of this photo "Wild Bill" was commuting from SoCal to drive the NorCal based* Cheetah *Top Fuel car.*

In 1967, Fresno, California-based Junior Fuel racer Gerald Enoch went Top Fuel racing with the Bad Guys digger. Unfortunately, their budget was limited and the Bad Guys never really became a Top Fuel force in California Top Fuel racing.

South San Francisco's Jessie Perkins parked his Runt *Top Fuel car and hit the NorCal quarter-mile with a longer car and Rich "Hippie" Bruckman at the controls. Bruckman proved you can't take a Top Fuel car off roading.*

John "Black Knight" Edmunds (far side) versus Tommy Walsh. The Crossley/Edmunds Top Fuel car was from Fresno, and the Hay, Haslum, and Walsh Wailer was from Hayward. The "Black Knight" took this race in 1965.

Rich Zoucha dropped by with his Californian *Top Fuel car and I shot a feature for* All American Drags *magazine in 1967. A neighbor stopped over during the shoot so I photographed her with the car.*

One of the first retro Top Fuel cars to hit the asphalt belonged to Chastain and Hedge, circa 1992. This was an all-new replica of an old style slingshot dragster.

Master painter and racer the late Tony Del Rio really did well on Ed Gonzales' show and go A/FD. The car won a lot at Fremont and Half Moon Bay. It also won at ISCA car shows in Oakland and San Mateo.

After his first Top Fuel ride in the Broderick, Blessing, and Swansboro blown fuel Chevy, Frank Bradley jumped into this nitro-burning Hemi digger with the team of Broderick, Bradley, and Jorgensen. This team had more success in Top Fuel and Bradley earned a new nickname, "Bucket Mouth," as he became one of the first trash-talking drivers in NorCal. Note: the author's movie camera is mounted to the top of the dragster.

The only DeSoto Hemi-powered Top Fuel car in NorCal was owned by Davidson, Pike, and Silvera. The Toy Top Fuel car was driven by Ron Davidson, who was a regular at Fremont in the 1960s. Davidson frequently lit the rear tires on fire, and the fans loved it. That is photographer Ted Stewart getting an earful of the Toy.

The Sears Point Raceway opener in 1969 was memorable for Junior Fuel racer Gene Gilmore. Gilmore had just passed the finish line when his fuel tank left his Bings Speed Shop-sponsored Chevy-powered Top Fuel car. No harm, no foul—a new tank was quickly installed and Gene was again off to the races. A long-time competitor in the Junior Fuel class, Gilmore first appeared in the mid 1960s with a Logghe "slot racer"-style Top Fuel car.

Here is a late 1960s battle in Top Fuel with two SoCal racers, Tommy Ivo and the **Howard Cams Special***, meeting in the NorCal Top Fuel Eliminator. The Howard Cams Special beat Ivo on a holeshot on this run.*

1960s Top Gas and Top Fuel driver Gary "Tango" Matranga, was driving the great Matranga, Hummeland, and Wills retro Top Fuel dragster in the early 1990s. Matranga drove Jim Guy's AA/GD, The Timmons Brothers Top Fuel Chevy, and the Cheetah II *Top Fuel car. Then he had a Camaro Funny Car and the last fuel car was a Vega funny car.*

Frankie Silva was the pilot of John Zoucha's Gateway Chevy AA/GD in August 1965. The beautiful full-bodied digger was a regular at NorCal racetracks.

Local Junior Fuel racers Berry and McDaniel take on Top Gas legend George Boltoff in 1965 at Sacramento. The NorCal based Junior Fueler driven by Rance McDaniel outran Boltoff's Top Gasser at this race.

The late Marc Danekas stripped the body off the Tognotti Speed Shop-sponsored Bushwacker in '66 or '67 and then replaced driver Jack Martin with himself. With Danekas at the controls, it meant wheelstands, more wheelstands, and a lot of Kodak moments for me.

My ol' buddy Rob Sterling ran an injected Hemi dragster out of Salinas, California. When I lived in Salinas, Sterling gave me a ride to the races. Whether I was in the bed of Sterling's pickup truck or riding in the cab, I had to get there. Sterling let me letter my name on his car, making it the first moving ad I had at the drag races.

DRAG WORLD

VOL. 3, NO. 49 FRIDAY, MARCH 15, 1968

FEATURE INDEX

MARTINEZ WINS TOP FUEL RACE

The Drag World *headline was correct: Frank "World's Fastest Mexican" Martinez did win in Top Fuel, but they forgot to tell the whole story. In the final, Martinez was racing Rich Zoucha, driving* The Wailer. *Both cars staged, the green light flashed, and away they went. The win light lit in Martinez's lane, but in the other lane, Zoucha was getting the hot-oil bath from hell. Unable to see, Zoucha veered into Martinez's lane and into his car. The two cars became one, leaving the Fremont asphalt and tumbling in the soft grass on the right side of the track. Both drivers scrambled from their destroyed diggers. Nobody was injured, but both cars were totaled. Martinez rebuilt his Top Fuel car, but the team of Hay, Haslem, and Walsh called it quits.*

The Whiz Kids Top Fuel car leaves first, but John "The Sutan" Batto will catch and pass it for a 1966 victory in the Eliminators at Fremont.

My first sighting of Don "Mad Dog" Cook was when he partnered with Dwight "Hay" Bale in early 1965. They took turns driving their Top Fuel car, and I think it was Cook at the wheel in this photo. Bale's helmet always had a visor.

Gerry Steiner always had the best-looking pit crew. His Rumchargers A/FD had Bay Area go-go girls on duty in 1965.

Old Top Fuel drivers never fade away — they come back to drive retro fuelers in the early 1990s. NorCal's Bill Dunlap was one that refused to fade. He just loves to go fast!

In the late 1960s, Kenny Safford and Ted Gotelli campaigned this Woody Gilmore-built digger, powered by a Gotelli-built 392-ci Hemi engine, throughout California.

Northern California had its share of show-and-go Top Fuel cars. One of those in the late 1960s was Dennis Hollingsworth's Dennis the Menace *full-bodied Top Fuel car driven by Jerry Vargas. Keeping their racing as a hobby, the duo ran in NorCal when business allowed.*

A hot-oil bath in the face and it's an off-road trip for Bob Meyer. The Speedmasters Top Fuel car really got bent up. Luckily Meyer walked away unhurt but very oily. This happened in the wide-open spaces of Sacto's shutdown area.

The husband-and-wife team of Walt and Debbie Anderson owned the East Side Kids blown Chevy Top Fuel car. The car was based in Northern California and could be seen on the quarter-mile in the mid 1960s. When he wasn't driving a jet car, J.D. Zink drove the candy-green Chevy, as did Don Clar.

He was always the old guy in the NorCal Top Fuel wars, but early in the 1990s, 62-year-old Don Argee debuted his new retro Top Fuel car. Argee was a veteran of many Top Fuel rides and one Fuel Roadster ride.

Early 1960s owner/driver Rich Guasco drove a 392-ci blown Chrysler Top Fuel car. The rear end spun out and Rich was severely injured, prompting him to sell the car to Chuck Flores. He went on to gain fame as the owner of the awesome Pure Hell AA/FA.

In Northern California, Steve Raynor ran one of those slingshots with a roadster body. The San Leandro racer campaigned this very clean, great-running racecar in the B/Fuel dragster class.

In January 1966, Jim Davis and his Texas Lizard *Top Fuel car cranked out a 209-mph run at his home track of Fremont. Davis could race with the best Top Fuel cars in the country.*

A typical pairing in Top Eliminator, 1965-style, was the Top Gas dragster versus Junior Fuel dragster. Arigoni, Gregg, Bernasconi, and Gewertz (near side) took on Nick Mura's Plymouth-powered Junior Fueler.

Willie Kinner and Gary Ritter were tough guys from Hayward, California, and Blood, Sweat and Nitro *was the name of their Top Fuel car. This scrappy duo hung with the best in the mid and late 1960s.* Blood, Sweat and Nitro *was Kinner tuned and Ritter shoed. When they got together, Ritter had to get his Top Fuel license. On his third and final run for his license, he blew the blower off the car and was knocked out when part of the back plate hit him in the head. He received his license once he got the car stopped.*

Bakersfield 1968: Frank Pitts had just purchased the Lizard *Top Fuel car from his former partners, Herbert and Bishop. Trying to fire up a cold Top Fuel car can be hazardous to your health On push down, the car made a turn for the Bakersfield popcorn stand, and crunch went the* Lizard.

Your humble author tried doing some race car illustrations. What's the verdict?

Fresno, California, is well known for its raisins, and in drag racing circles it was also home base for the show and go Top Fuel dragster of the Berry Brothers and Hughes, a very popular race team in the late 1960s. The first car I saw run by the Berry Brothers was a Chevy-powered roadster in 1963–64, and the next thing I knew they were running a Top Fuel car with Claude "the Wasp" Stark at the controls. Butch Maas also spent some weekends behind the wheel. When Junior Fuel racer Dwight Hughes was given the chance the team clicked.

In Bakersfield 1969, John "The Zookeeper" Mulligan sent the blower on the Beebe and Mulligan Top Fuel car into orbit. He got the car stopped safely and exited with some nasty burns around his eyes from the flash fire just prior to the big boom. When the burns started to heal, Mulligan looked a lot like a raccoon.

A regular on the NorCal Top Fuel circuit was Chuck Flores, a longtime racer who raced the old "Sour Sisters" Shark *car in the mid 1960s. Either Flores or Rich Bruckman could be seen behind the wheel of the Shark.*

Rich "The Hippie" Bruckman drove Jesse Perkins' Runt *and* Cow Palace *Shell Top Fuel cars as well as Chuck Flores'* Shark *in the mid 1960s. Perkins and Bruckman became a mainstay in Top fuel racing and in the 1960s it seemed like every Top Fuel Show had Perkins* Cow Palace *Shell Top Fuel car in it. Chuck Flores was a fixture in NorCal Top Fuel racing starting in the early 1960s.*

Touring the B&N Automotive way in 1967. Owner Noel Black always carried a spare.

There was some great Top Fuel racing in Lodi, California, during the mid-1960s, and like most tracks in NorCal, Lodi had no guardrails. Photographers were on their own when shooting action shots. Don Cook tried to put his Top Fuel car's front wheel in my pocket one Sunday afternoon. He missed by a couple of feet. If you look at his tire tracks, you will see where I had been standing. Me 1, Cook 0.

Southern California had its **Freight Train***, and Northern California had its own hard-charging twin-engine Chevy digger. Newark's Leo Dunn had one of the best-running twin-engine diggers in the USA. Teamed with Rich Burnelli, owner of Central Chevrolet, Dunn made his mark in NHRA Division 7 WCS Series races in the late 1960s.*

In the good ol' days you could find the Galli Brothers' Olds-powered A/FD at local NorCal dragstrips. When the Olds proved outdated, a 392 Hemi took its place, and driver Rich Galli put the car into many a winner's circle.

What do AA/FA drivers do to relax in the late 1960s? Dale "The Snail" Emery could be found in Mike Fuller's Top Gas dragster when Pure Hell was parked at its NorCal home of Pleasanton. Emery is no stranger to Top Fuel or dragsters as he drove the K&P Olds Top Fuel car in the early 1960s.

The late Frank Martinez was a bucks-down Top Fuel racer with a lot of heart. Early in his career, he teamed with Dave Uyehara on The Hustler *Junior Fuel dragster. Martinez jumped to the Top Fuel class by driving* The Wailer, *then racing his own Leroy Jorgensen-powered Top Fuel car. Martinez was a family man who always had his daughters at the races with him.*

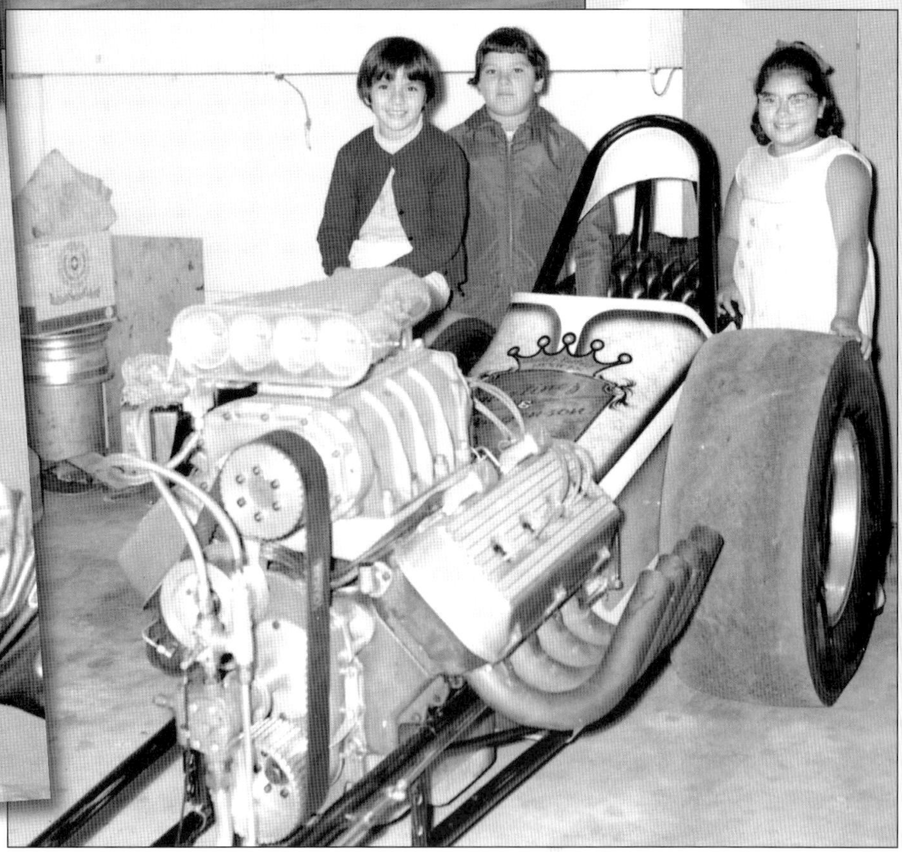

Veteran driver Dave Uyehara not only drove a nitro-burning race-car, but was also a chassis builder. Known as the "World's Oldest Living Kamikaze Pilot," Uyehara started a Cadillac-powered Model A Altered. He then teamed with Frank Martinez and raced The Hustler. *Next was Top Gas and then Top Fuel. Uyehara was a fearless Top Fuel driver who tested his chassis the hard way, and since Dave is still here, I guess he builds a great product.*

Two of the heavyweights in northern California's Top Fuel class faced off at Sacramento Raceway in the late 1960s. Walnut Creek carpet man Dennis Baca (near side) went wheel-to-wheel with the Hayward-based team of Willie Kinner and Gary Ritter.

Bob Keith in the Goodknight, Keith, and Williamson Chevy-powered Top Gas digger (far side) led a wheelstanding Bill Dunlap in Jon Halstead's Hemi-powered digger. Both teams were from San Jose and there was a big rivalry between the two.

The last drag race for the Arigoni, Gregg, and Bernasconi Top Gas Dragster was at Vacaville, California, in the mid 1960s. A blown engine at the finish line, a ball of fire, and a couple of endos spelled the end of the gas digger. A strong rollcage saved the driver. They rebuilt and returned to the quarter-mile Top Gas wars.

Looks like some of those gallopin' hinges fell out of the 392 Hemi, which was very costly and messy for driver Rich Zoucha.

In 1967, Jim Nicoll and Don Cook ran the Der Wienerschnitzel *Top Fuel car. The following year Jim paired with old buddy Leroy Goldstein, and the* Der Wienerschnitzel 2 *was born. Nicoll finished the 1960s and early 1970s on his own as driver and owner of the* Der Wienerschnitzel.

A late-1965 in Top Fuel at Fremont Drag Strip between the big bad 392 Hemi Top Fuel car of Jim Davis (far side) and George Santos' injected nitro-burning Chevy-powered Goldfinger. *When the green flag dropped, Santos' driver Jim Perry was off and running with Davis and his Hemi digger in hot pursuit.*

It was the late 1960s and Frank Martinez was in trouble with his Top Fuel dragster. Martinez had just blown the front of his nitro-burning Hemi all over the Fremont finish line. The car was on fire and out of control and Martinez had received flash burns in his eyes. Thankfully, the car stayed on the racetrack and stopped safely. Martinez suffered burns on his hands and face, but he wasn't the only person hurt. Photographer Mike Bagnod suffered a chipped elbow when he was hit by flying block pieces.

Never heard of Top Fuel owner/driver Don "the Golden Shoe" Petrich? He was a low-buck Top Fuel racer in the mid 1960s and the original driver of Rich Guasco's Pure Hell *Fuel Altered. He wore one gold shoe when he drove.*

A lot of used SoCal Top Fuel cars ended their days with NorCal owners. One of these owners was Jim Lind with his Up Tight, Out of Sight *Top Fuel car that was formerly the Winkel, Trapp, and Glenn* Magicar. *Lind didn't make any Top Fuel records disappear, but he had a lot of fun during the late 1960s.*

A very young David Baca was his dad's official VHT pourer in the early 1970s. Back then, David was very proud of his Top Fuel driving dad, Dennis. These days, David is a 330 mph, 4-second Top Fuel driver/owner, and Dennis can be proud of his son.

Salinas Drag Strip circa 1966 was a bumpy old air strip with a flagman and a cow pasture on one side, but it worked.

A great starting line launch and then "boom" on the Lions Drag Strip. This unknown SoCal Top Fuel car sheds the blower about 100 ft off the starting line with parts and pieces flying everywhere but nobody was hurt this time.

Bill Wishart was in a wheelchair as a result of a shooting accident when he was a kid, but that didn't keep him from going Top Fuel racing. I met Wishart in the mid 1960s, and he kept telling me, "I'm going Top Fuel racing." One evening, when we had too many beers, we decided to see how fast his wheelchair would go across the parking lot. So with me pushing, we tried to set a new land speed record for wheelchairs in Fremont. We were really cookin' across the lot until we hit a small rock, and we ended with a flip and a crash. I went over to Wishart who had, of course, departed from his chair. I helped Bill back into his chair and we found out later he had broken his toe but couldn't feel it at the time. Wishart's dream came true at the beginning of the 1970s. The late Pete Ogden built a super-light chassis and everything started to come together with Bill Alexander as the driver. If you want something badly enough, dreams can come true.

The Fremont fans loved their Top Fuel racing, and they packed Fremont to watch their nitro heroes do combat in the late 1960s.

When Rich Bruckman vacated the Jesse Perkins Cow Palace *Shell Top Fuel ride, Bobby Hightower filled the vacancy. The Southern California journeyman driver did well for Perkins.*

This was state-of-the-art in NorCal trailers during the mid 1960s: John Halstead and his Western Manufacturing Special *windowed racecar hauler.*

The "Fresno Flash," Johnny Edmunds, teamed with the legendary Tony Waters in the mid 1960s. The duo was a threat to all they competed with throughout California. John "the Black Knight" Edmunds took no prisoners with the potent brown Top Fuel car.

TOP FUEL IN SOUTHERN CALIFORNIA

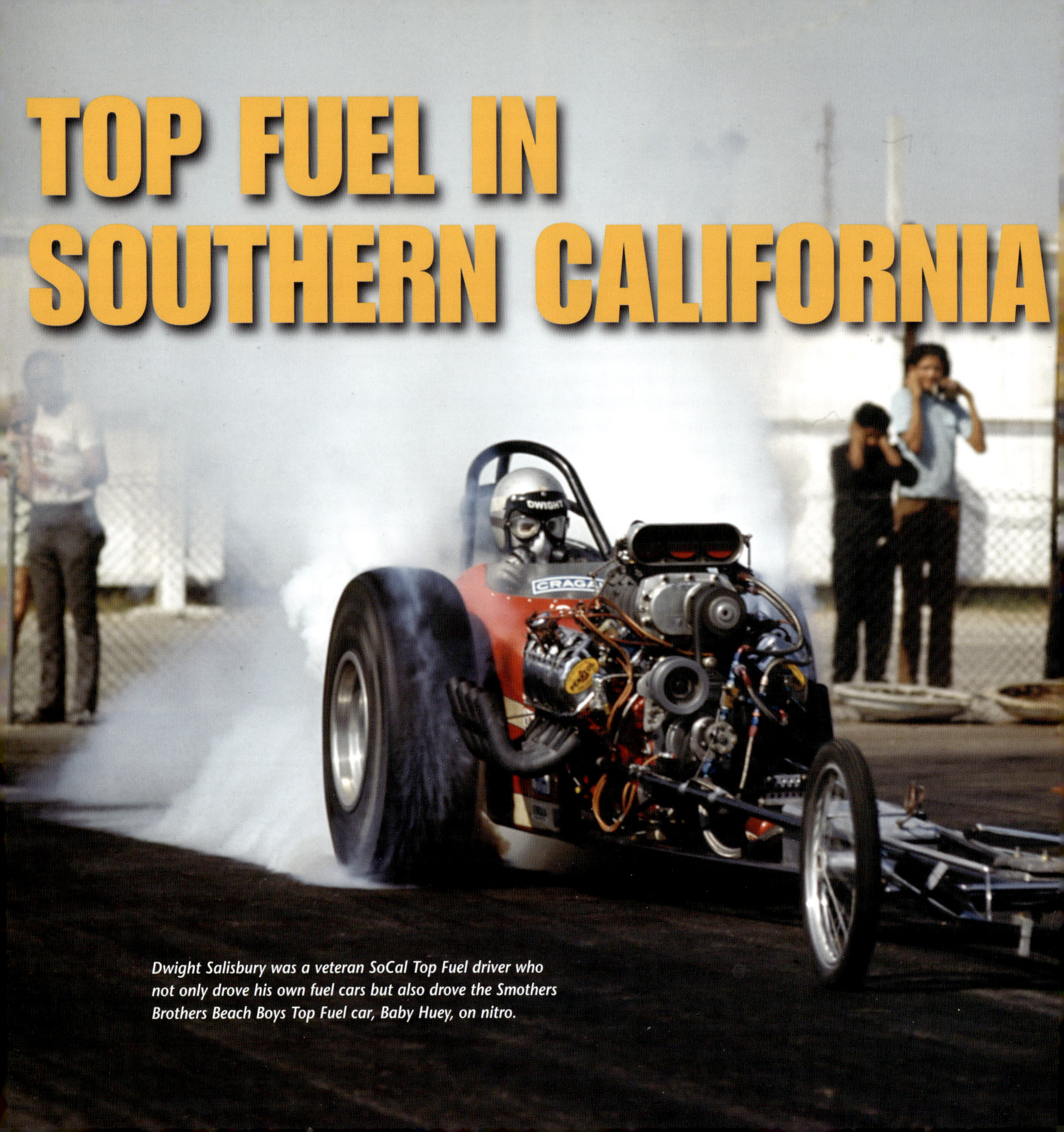

Dwight Salisbury was a veteran SoCal Top Fuel driver who not only drove his own fuel cars but also drove the Smothers Brothers Beach Boys Top Fuel car, Baby Huey, on nitro.

From 1965 to 1970, I think I wore a path from NorCal to SoCal to cover the country's best Top Fuel racing. One of the first races I photographed in Southern California was at Irwindale Raceway, which was a replacement for San Gabriel, a racer favorite that closed in 1963. Irwindale was located next to rock quarries and next to the then unfinished I-210 freeway, right above Azusa. It was smoggy, dusty, and sandy, but the track was well cared for and had great food and grandstands for the SoCal drag fans. One could find a vast array of Top Fuel stars there on any given weekend.

My biggest problem with Irwindale was management. Even though I was being published in magazines and drag papers and had six different press passes, they demanded photographers provide them six 8 x 10s of cars racing at their racetrack. Since they didn't say which cars they wanted photos of, I shot pictures of the ugliest stock cars going down the track. I delivered the photos, and did they bitch. So the next time I printed photos of the Top Fuel cars but used a very weak fixer on the prints, which caused the photos to fade. I went back and forth with the management for a while but when I became the NHRA Division 7 photographer, the harassment finally stopped.

I have another amusing story involving one of Irwindale's managers. One half of the Irwindale management was a civic leader in the area—let's call him Richard. It seemed Richard had made enemies in local political circles, and one of those enemies needed Richard's vote on a pending bill. Knowing they were unlikely to get the deciding vote, they hatched a plan to get it and invited Richard to a Mexican restaurant for a peace-making meal. A great dinner was served and Richard downed all of his enchiladas. As the evening wore on, Richard started to feel lightheaded and then passed out. His foes quickly carried him from the restaurant to a waiting car and whisked him away to the nearest "no-tell" motel. Waiting were two nude hookers, and let's just say lots of very incriminating photos were taken while Richard was dead to the world. The culprits dressed Richard and returned him to his car in the parking lot of the restaurant. You can guess what was next. If Richard didn't vote the right way, the photos would go public, and because he was a family man and pillar of the community, the photos would destroy him. What they didn't count on was Richard's courage. He turned his extortionists in to the local police. They went to jail and Richard's reputation didn't suffer. That was the end of the case of the spiked enchiladas.

Another "action" tale associated with Irwindale involved yours truly, my photographer pal Mike Bagnod, Eves Tallchief

Southern California's Dave Mackenzie traveled to Northern California to race the Reath and Herbert Top Fuel car. Mackenzie went head to head with the best fuelers in NorCal. This beautiful black race car won its share of gold.

(the ex-flag starter at Fremont), and Billy "the Stick," a drag fan and pool player. On one of my ventures to Irwindale, we needed to find a ride to the races since none of us had a car big enough. We called in a favor from Ed Carter, a Funny Car driver, partner in the *Chevy 2 Heavy* Nova, and top-notch salesman at Central Chevrolet in Fremont. Carter was able to put us into a brand-new demo Chevy for the weekend, so early on Saturday morning we picked up our shiny new Chevy and off we went. A quick six-and-a-half hours later, we were at the gates of the 'dale.

It was the Grand Prix of drag racing—a giant Irwindale event with lots of action and rolls of film shot. After a long drive down and a long day and night of drag racing, we piled into our loaner and headed home. Down the 605 freeway we went, and I soon got the very bright idea to throw a cherry bomb out onto the freeway. I enjoyed hearing cherry bombs go *boom,* and I had one just burning a hole in my pocket. Billy was driving, Chief was riding shotgun, and Bagnod and I were in the back. We were running about 65 or 70 mph down the freeway, and I asked Bagnod if his window was down. He held his arm up and motioned towards the window. I asked Chief for the car lighter and he handed it back to me. I lit the cherry bomb and threw it toward Bagnod's window. The only problem was that his window was up, not down, and the cherry bomb hit the window and fell at Bagnod's feet. I saw this and went into a fetal position. Bagnod turned white and in one motion was in my lap. The cherry bomb exploded and the world ended within the loaner car. Smoke and carpet fibers filled the Chevy. Bill and Chief of course were totally unaware of what I had done. Bill got the smoke-filled Chevy stopped on the freeway shoulder and we all dove out of the car. None of us could hear and we had carpet fibers in our hair and clothes. The bomb had blown out the dome light, and pieces of plastic were everywhere. After clearing the car of its smoke, we climbed back in and continued our now very quiet journey back home. Carter was none too happy about his demo when we returned it, but he didn't make me pay for the new carpet, dome light, or interior cleaning. However, he never again loaned us a car, and lucky for us, our hearing returned a few days later.

Not all Irwindale stories had happy endings. At Irwindale in 1966, SoCal lost one of its favorite Top Fuel drivers, Allen "Lefty" Mudersbach, who made his last pass at this track. At the finish line, the front tire blew and a large piece struck Lefty in the face. The car left the track and ended up in the rock quarry on the right side of the shutdown area. Mudersbach was fatally injured, and his death was a great loss to Top Fuel racing in SoCal.

Back to a lighthearted tale of Top Fuel nonsense. How about three Top Fuel cars running for the final? It happened at Irwindale. Late one Saturday night in the mid-1960s, John Harbart's *Pegasus* Top Fuel car was to race another SoCal Top Fuel car for Top Eliminator. As the time drew close for the two racers to meet, it looked bleak for the *Pegasus*. The Harbart car had hurt the engine in the semis and a frantic repair was taking place. If *Pegasus* couldn't make the call, an alternate would replace it. The call was made for the two cars and it was time to race for all the money. Unfortunately, *Pegasus* was nowhere to be found. The opponent and the alternate started pushing down from Irwindale's upper buildup area. Both roared to life and the fans cheered. But wait, there was a third Top Fuel car coming down the pushdown road: It was the *Pegasus*.

In 1966, Blake Hill's new fueler was an exact clone of Roland Leong's* Hawaiian. *Hill's* Grecian *was exact down to the paint and letter style. That is where the cloning ended, because Hill didn't have Leong's tuning or Mike Snively's driving abilities.

It was May 1965 when I saw one of the most beautiful Top Fuel dragsters of that time. It belonged to Larry Stellings and Bobby Tapia, a SoCal duo who were in Sacramento for a big Top Fuel race. Tapia, the "computerized" driver, did the driving on the candy tangerine metal flake car.

Picture this: Two Top Fuel cars lined up to race for Top Eliminator, and a third tried to line up also, but of course only two can run down the track. The starter told *Pegasus* to shut off, but the driver refused. An Irwindale employee stood in front of the disobedient driver. The *Pegasus* pulled around the employee and staged alongside the other two running cars; the only difference was that *Pegasus* was on the fire-up road and not the drag strip. The light turned green and three Top Fuel cars smoked their tires. The *Pegasus* did shut off, but not before scaring the heck out of the Irwindale personnel and the drivers of the push trucks on the road. Harbart and his *Pegasus* were banned forever from SoCal drag strips, but strangely, about a year later a Top Fuel car showed up looking like the *Pegasus* with a driver named Cliff Roa, who looked a lot like Harbart.

If you loved Top Fuel cars, the Mickey Thompson 200 mph Club races in 1965 and 1966 were your cup of nitro. Held at Lions Drag Strip, they were the Mecca of the Top Fuel racing world, featuring Top Fuel, Top Gas, and Junior Fuel cars. When Thompson chose to forego the race in 1967, a new nitro promoter took

One of the most popular Top Fuel drivers of the mid 1960s was Tom "The Mongoose" McEwen, who spent a lot of time behind the wheel of Lou Baney's Yeakel Plymouth Top Fuel car. For a while, "The Mongoose" seemed unbeatable in the purple Top Fuel car.

Ron Rivero hits the Irwindale quarter-mile in 1966 driving the Frantic Four/K&G Associates Top Fuel car. What was to be a temporary driving job turned into a full-time ride for Rivero as the original driver, Norm Weekly, had lost his desire to drive full-time after the death of his good friend Denny Milani.

California on a great run of dragster racing with the Professional Dragster Association (PDA). An expert at building aluminum race bodies and a big fan of nitro drag racing, Doug Kruse wanted to have the biggest and best Top Fuel race in the USA. He chose the drag strips of California to hold a series of incredible Top Fuel races. From 1968 to 1975, Kruse held his PDA races at Lions, OCIR, Fremont, and Sacramento, and only the finest racers and a "Who's Who" of racing attended. In my opinion, it was the best group of drivers ever assembled on the quarter-mile. In those eight years, they were my favorite races to attend and photograph.

My favorite drag strip in the Los Angeles area was Lions, and of course, Lions had its share of sad and happy stories. The state of Maine had one Top Fuel owner/driver, Joe Jackson, and in the mid-1960s, Jackson towed to California to compete at Lions. Unfortunately, his first pass down the drag strip was also his last when a tire-smoking run in his Maine Auto Parts digger landed him into a Lions light pole and claimed his life.

Lions had some of the most horrendous-looking Top Fuel crashes ever, and most of the drivers walked away. One was a young Hawaiian kid named Roland Leong, owner of the famed *Hawaiian* Top Fuel dragsters and Funny Cars. Leong ventured to Lions in the early 1960s to be a stud Top Fuel driver. His driving career didn't last too long in the new Kent Fuller *Hawaiian* chassis as a banzai pass down the Lions quarter-mile and a trip through the "end" sign posted at the 1,320 mark put an end to it. Sans one very bent race car and with a very sore body, Leong decided that tuning was the best place for him.

A couple of veteran Top Fuel drivers had their own problems at Long Beach, including Don Moody, whose steering broke at speed, taking him on the ride of his life in the early 1960s. Emergency crews expected the worst when they went to the crash site and found Moody and just Moody. No race car, no rollcage, just Don Moody. He had an injured toe and bumps and bruises. What a lucky, lucky guy.

When two crazy motorheads got together to go Top Fuel racing, the result was the Yellow Fang. *In the mid 1960s George "The Bushmaster" Schreiber convinced his longtime friend Ed "Big Daddy" Roth that Top Fuel racing was where it was at. Schreiber's idea was to introduce a Roth creation into the SoCal Top Fuel class. All Schreiber needed was money because he already had Roth's blessing. Schreiber's idea of gathering cash for racing was unique, but it worked. Every day when Roth came to his office/shop in Maywood, he checked his pockets for change and tossed all the change up on the roof of his office in the Maywood shop. Schreiber knew of Roth's change habit, so when it came time to finance the Top Fuel car, George asked Roth if he could gather the change on the office roof. "Sure," said Roth. Schreiber began to gather coins and more coins. There was enough to purchase a Jim Davis chassis and get a good start on many other items. Master designer Steve Swaja designed the* Fang*, and Bill Demerest built the 392 Hemi. This show-and-go Top Fuel car ran a limited amount of races in the mid-to-late '60s. The car is still around and the spirit of Ed Roth is still alive in "Big Daddy" Don Garlits' Museum of Drag Racing in Ocala, Florida.*

Moody had crashed at night, but Rick "the Iceman" Stewart wanted to make sure everyone saw the action when he crashed during the day. Stewart took the two-week-old Adams and Wayre Top Fuel car on a wild ride on August 14, 1965, and received an oil bath at 207 mph, which caused the car to veer and crash on the right side of the Lions shutdown area. When all the dust had settled, the new Top Fuel car was junk and Stewart was beat up but able to walk away. The Adams and Wayre team returned three weeks later with a new Top Fuel car and John "the Zookeeper" Mulligan at the controls.

John Collins' crash in the Dave MacKenzie Top Fuel car was avoidable, but memorable. It was late and dewy at Lions, and Collins and MacKenzie were in the final for Top Fuel Eliminator. Their competition was a no-show, so Collins suited up and buckled into the sleek black Top Fuel car. Onto the Lions roller starter went the nitro-powered digger. The rollers turned, and the 392-ci nitro-gulping Hemi came to life. Raw nitro filled the starting-line area as the Hemi warmed up. Collins got the signal from owner McKenzie that it was time to stage the car. A single for Top Fuel at Lions was a rare happening, but Collins staged and starter Larry Sutton hit the switch and the green flashed. In my camera's viewfinder I saw Collins' arm out of the car flashing a V for victory as he left the starting line, and he came by the row of photographers with his arm still high in

The La Mirada, California, fireman Jim Dunn climbs from the Dunn and Reath Top Fuel car. Dunn had just completed a successful pass down Lions Drag Strip in 1969.

the air. Then the clutch came in and the front wheels headed for the stars. Up, up, and over onto the guardrail went the car. It seemed like forever that it rolled down the Lions guardrail. Before the emergency crew arrived, Collins was out of the very bent race car, uninjured but very embarrassed. If you ever want to get John Collins mad, just flash him the V sign and then run.

One more bit of craziness at Lions was toward the end of the 1960s when Mike Kuhl and Billy Tidwell debuted their all-new unpainted Top Fuel car. After the required shakedown runs, Tidwell put the new car into the Top Fuel Eliminator field and surprised everyone when he advanced to the semifinals. As the semi-finals were ready to rock, the cars were on the rollers and the sweet cackle of nitro-burning Hemis filled the Long Beach night. Tidwell carefully staged in the left lane, the green light flashed, and both cars headed to the end of the 1,320. With huge header flames, both cars were even at the 1,000-foot mark when all went wrong for the Kuhl & Tidwell Top Fuel car. The 392-ci Hemi gave up its crankshaft through the bottom of the oil pan, and the car looked like an earth-bound meteor. The fireball went up on the guardrail and over it. From the starting line the car looked like well-done toast and everyone feared the worst, but Tidwell walked away from the very well-built Kuhl and Tidwell Woody-chassis car. The new car got used up in just one day of racing, but Tidwell was around to race another day.

The Top Fuel car from Rhode Island is Jimmy King and Donnie Marshall's* El Diablo. *King did the driving and Marshall did the tuning.

Just south of Los Angeles was the new kid on the block, Orange County International Raceway. OCIR opened in 1967 and was state-of-the-art as drag strips go, with a price tag of more than $1 million to build. First class was how manager Mike Jones and company ran OCIR, and it was drag racing's crown jewel. Unfortunately, OCIR finished its first year of operation on a very sorrowful note. On New Year's Eve, 1967, Mike Sorokin of *The Surfers* fame was driving for Tony Waters when a massive clutch explosion at the finish line took Sorokin's life. The little guy with a big smile and big heart was gone — another loss for drag racing.

I had a lot of fun at OCIR. The racing was great and it had a great place to photograph cars on and off the track. I don't think Mike Jones liked me much, but he put up with me because I

A staple in Top Fuel racing in California in the mid 1960s was Bob Mayer and the Speedmasters Top Fuel car. The team was based in the San Diego area and could be seen at racetracks throughout California. Mayer did some great wheelstands with the digger, actually grinding holes into the driver's seat. He may not have been a great racer but he sure did put on a great show. I guess if you can't win, it helps to look good.

was the most published drag-race photographer at the time. I am sure that burning a hole in the OCIR bleach box probably didn't win me any points with him. I guess the 5 gal of gas we used was a bit too much for a fire burnout. If Jones had known I photographed a nude girl on top of the OCIR tower during a race, he probably would have banned me forever. Yes, Tim Marshall and I got bored waiting between rounds, so we photographed a girl on top of the tower while stock cars did burnouts in the background.

One of the drivers I was fortunate to get to know was Carl Olson. We hit it off, and I became Olson's unofficial photographer and press agent. At the time, Olson was racing an ex-show dragster. His *Hooligan* injected Plymouth-powered digger won its share, but Olson yearned to go really fast. Besides racing his own car, he was a crew member on the Jack Ewell's Top Fuel team. Whenever Olson crewed for Ewell, my phone rang and Olson invited me to travel to Southern California.

Ewell was a longtime dragster racer, but in the late 1960s it was time for him to hang up the fire suit and let one of the young lions of SoCal racing take over. Ewell's choice was a very talented Junior Fuel driver, Ed Allison, who drove the Allison, Crowe, & McCarroll Junior Fueler with a nitro-injected small-block Chevy that was one of the great cars in the class. However, Ewell's plan was derailed when Allison suffered a fatal crash in Carlsbad, California. During a run with the Junior Fuel car, the clutch exploded, tearing the car into many pieces. Allison barely survived the explosion and crash and laid in a coma for almost three months before passing away. SoCal had lost another popular driver, and Ewell had lost a friend. Ewell looked no further than his own pit crew for a replacement, and Carl Olson finally got his chance to go really fast. Under the watchful eye of the "Ol' Gray Fox," Olson got his Top Fuel license at Lions. He made Ewell proud and became one of the best Top Fuel drivers in the country.

Rick "The Iceman" Stewart didn't just appear from nowhere to become the starter for the NHRA POWERade Drag Racing Series. He drove Top Gas and Top Fuel with Gene Adams and Jack Wayre, as well as with Bill Crossley. "The Iceman" was one heck of a shoe in the 1960s.

The late Steve Carbone debuted his sleek, new Top Fuel car in style at the Ontario Motor Speedway Grand Opening ceremonies in August 1970, providing the media with smoky burnouts. Of course, Carbone was a veteran driver of many a Top Fuel car including John Bateman's Atlas Oil Tool Special.

Journeyman Top Fuel shoe "Captain Billy" Tidwell, circa 1970. Tidwell drove for Kuhl & Tidwell, the Skinner Brothers, Lou Baney's Folger Ford Top Fuel car, as well as the Lisa & Rossi Top Fuel car.

Leonard Abbot and Joe Lee ran the Shifter Top Fuel car out of San Diego. The Shifter was a rolling lab for Abbott's Lenco Transmission business.

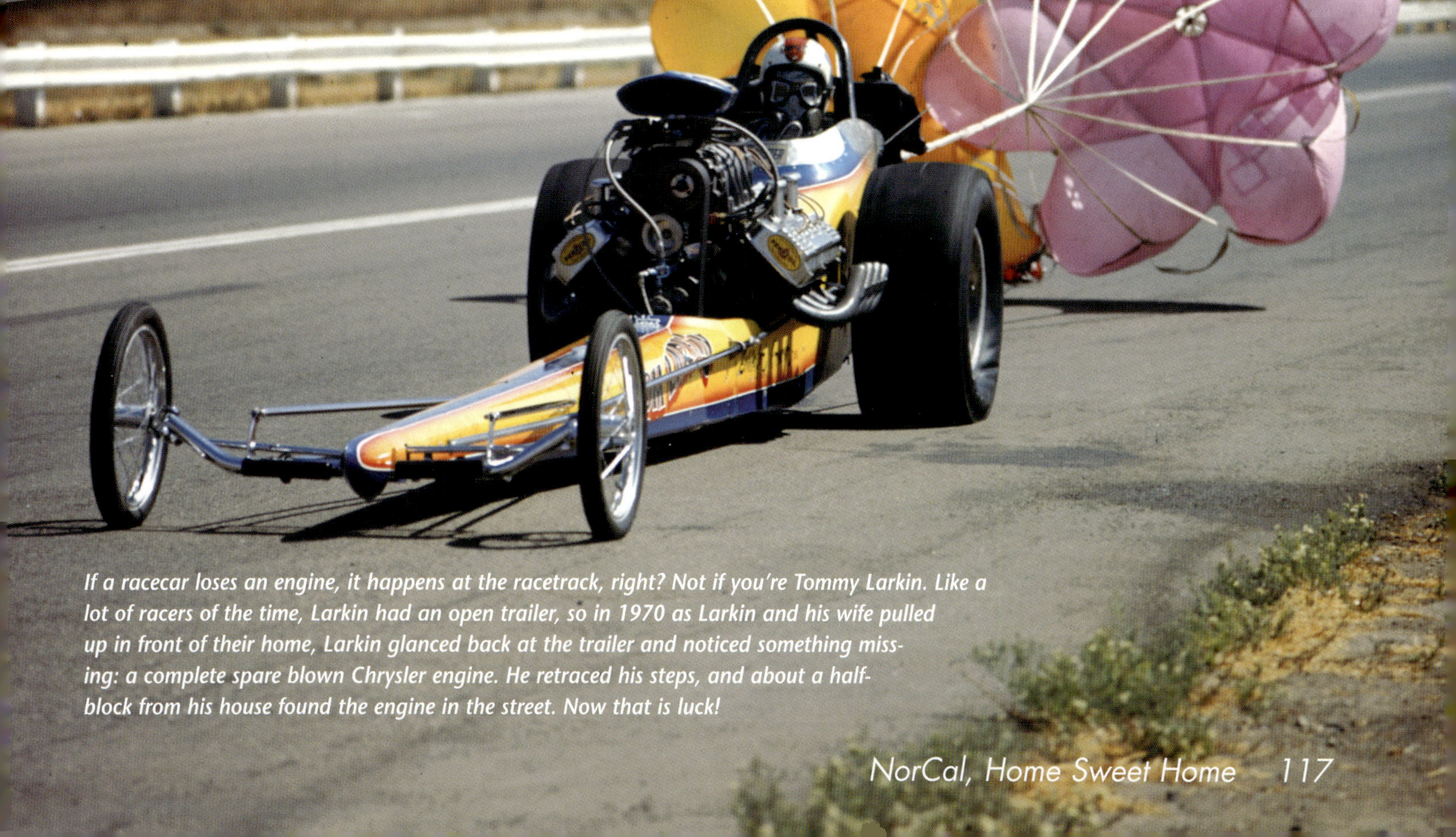

If a racecar loses an engine, it happens at the racetrack, right? Not if you're Tommy Larkin. Like a lot of racers of the time, Larkin had an open trailer, so in 1970 as Larkin and his wife pulled up in front of their home, Larkin glanced back at the trailer and noticed something missing: a complete spare blown Chrysler engine. He retraced his steps, and about a half-block from his house found the engine in the street. Now that is luck!

This photo of Jack Ewell and crew was taken at Pomona in 1967. The "Ol' Gray Fox" (Ewell) was one heck of a racer. Driving his Ardun-powered dragster, he won Top Eliminator at Santa Ana on June 26, 1955. He beat Calvin Rice, Art Chrisman, and the Bean Bandits on that long-ago day in drag racing. The crewmember to the far right is soon-to-be Top Fuel stud driver Carl Olson.

In April 1969, one of the most popular Top Fuel drivers in Southern California, "Jumpin' Jeep" Hampshire, survived one of the worst crashes ever at OCIR. At more than 200 mph, Hampshire hit the guardrail at the finish line. In the ensuing mayhem, he suffered seven broken ribs, a punctured lung, both arms broken, both wrists broken, both collarbones broken, and a ruptured liver. Hampshire lived to tell the tale, but his driving days were over, and the Top Fuel class lost a very talented driver.

There was only one Gary Gabelich, Top Fuel driver, land-speed record holder, and all-around insane person. From Top Fuel dragsters to Top Fuel Hydro boats, if it went fast and was scary, Gabelich drove it. With his trademark purple plume and black facemask, he is seen in the Purple Gang Top Fuel car. The two Gabelich stories I most enjoy are when he went 60 mph on a skate-board on a Los Angeles freeway and the when he took the Valkyrie jet car into the Hollywood Hills and fired it up for his friends to see.

San Fernando Raceway manager "Hand Grenade" Harry Hibler drove anything with a fuel-burning Hemi in it and made his mark with many oildowns at numerous SoCal dragstrips. He and car owner/tuner John Smyser were runners-up to the bucks-up Tony Nancy for the Top Fuel title at Bakersfield in 1970. You gotta love a guy who paints "Hand Grenade" on his helmet and lives up to his nickname.

This awesome twin-engine fueler was The Syndicate *of Jack Raitt, Curt Caldwell, and Al Forthmann. Built in 1971, the Syndicate never lived up to expectations. A bit of sadness goes with this SoCal Top Fuel car, as driver Phil Ditmars, "the Boy Wonder," who was a young, good-looking surfer type, was stabbed to death in a senseless bar fight in Northern California; a true waste of a great person.*

The "Ridge Route Terrors," otherwise known as Warren, Coburn, and Miller, are slow to stop at Orange County International Raceway.

This was one of those Top Fuel cars I saw once at Lions in 1969. It has the name March Wind, *but the life of me, I cannot figure out who owned it or drove it.*

The late Pete Millar was not only a drag racer but also an artist. Mr. "Drag Cartoons" Millar could be found piloting his blown fuel-burning 289 Ford-powered Chicken Coupe *throughout California during the mid 1960s. The beautiful green coupe/dragster didn't set the world on fire, but Millar had fun. He even took the car to Sweden for a successful tour, after which he packed it into its shipping container for the trip home and headed for the airport. While Millar was flying home, the* Chicken Coupe *and all its spare parts and tools were dropped into the Swedish bay. A couple days later the container was raised from the bottom, the water drained, and it was put on a ship bound for the USA. A couple of weeks later, Millar was reunited with his race car, but unfortunately it wasn't the reunion he had expected. The shipping company had neglected to tell Millar that his container had played submarine. Upon opening the container, he discovered salt water does nasty things to racecars, parts, and tools. It was the end of the* Chicken Coupe *for there was no way to rebuild.*

Senior citizens rejoice—this one's for you! In Southern California during the mid 1960s, the over-55 crowd had a hero in Top Fuel named George "The Stone Age Man" Hutchinson, who raced the "youngsters" in his pink-and-white Top Fuel car. Hutchinson and his pink-plumed helmet were regulars at all the Southern California dragstrips. He didn't limit his driving to his Top Fuel car. The ancient one could be seen at the controls of Don Green's Rat Trap *AA/FA and Fling Traylor's* USA-1 *turbine dragster.*

Hanging out with Bill Holland and John Gudel Jr. proved to be a very interesting time. Holland and Gudel ran the Art Linkletter House Party Top Fuel car, and one Monday afternoon I went with photographer Les Lovett to their shop. It seems some "different" photos were needed. On hand was Lois, a nude dancer from up the street. The car was put in front of the shop and Lois did her thing, starting out with her swimsuit, and before we knew it, posing before God and everyone, sans top with the digger, out in the open and visible from the street. Les and I were clicking away, and all of a sudden we heard a crashing sound. We turned to see that a guy had run his car up the curb and hit a small light pole. So we decided that maybe we should finish the shoot inside the shop.

From the mid 1960s to the closing of Lions Drag Strip in 1972, L&M photos were sold at Lions. Larry "Max" Maxwell was L&M's photographer and owner. Some of the nicest color photos available to fans and racers came from Maxwell's camera. He didn't trust the drag-race media, so he only dealt with fans at Lions and a few racers. His display of photos at Lions were breathtaking. Max knew how to market to the fans at the dragstrip, too.

Danny "The Mangler" Ongais at the controls of Carl Casper's **Young American** *Top Fuel dragster, circa 1970. he looked good in Casper's Top fuel car but Ongais wasn't the easiest person to deal with. Gary Cochran took over the driving before Carl put it on the ISCA show circuit.*

Butters & Gerard racing Top Fuel at Lions in 1970.

Slowing down after a bit of an oil bath in the OCIR shutdown area is Mike Clancy in the Butters and Gerard Top Fuel car. He was okay and he will never rust.

Ron Manke made a pass down the Pomona quarter-mile in the Cerny and Lins fueler out of Minnesota. Manke went back to Minnesota and Don Moody came out of retirement to drive the Wes Cerny-tuned Top Fuel car which was now based out of Southern California.

When Dean Moon of Moon equipment wanted to go Top Fuel racing in the late 1960s and early 1970s, his choice was Jack McKay and his unknown driver Clayton Harris. The Moon Burst *Top Fuel car was born and could be found primarily on the AHRA Top Fuel circuit and a few major NHRA events. The team of McKay and Harris made their mark in Top Fuel with the* New Dimensions *rear-engine Top Fuel car a couple of years later.*

It was a NHRA World Championship Series race at Irwindale during the early 1970s, and first round of Top Fuel was just starting. I had wandered down the track for a two-car race photo with the crowd in the background when the first pair, with Tom Tolar driving the Tolbert and Tolar TNT Special *on my side, staged. The Christmas tree went green and here came the race cars. Just past the tree, a flame pushed out the near side of Tolar's engine followed by a huge fireball. Pieces were flying all over, along with a hot oil spray, so I decided it was time to duck behind the guardrail. The flaming Top Fuel car went past, and like a gopher peeking out of its hole, I was shooting photos again. When the fire, smoke, and pieces cleared, there was the* TNT *Hemi sitting on the racetrack. No car, just the engine, or what was left of it. I checked myself for holes and found none. Tom Tolar had made it to the finish line sans engine, and he was okay, too. Just another day at the races.*

Don Prudhomme's Wynn's Winder *was one bad hombre in Top Fuel. I remember seeing the yellow-and-red racer in the winner's circle many times in the late 1960s and early 1970s. "The Snake" thundered in this car no matter where he was racing.*

One of the first cars I photographed when I moved to Los Angeles in 1970 was Jimmy Inge's Sons of the Rising Sun *Junior Fuel car. Inge's Junior Fuel car ran as good as it looked and was based at his Shell service station in Inglewood.*

Kay Sissell's twin was a very cool car. The Los Angeles resident was a part-time racer and a full-time fireman, very strong in Competition Eliminator.

One of my personal favorites in Top Gas was the **Odd Couple***, which had one blown Hemi and one blown Chevy nestled in one chassis. The very capable Walt "I can drive anything" Stevens put the* **Odd Couple** *in the final round quite a few times.*

Bonneville regular and Top Fuel racer Roger "Ramjet" Gates blew the oil pan off his **Cracklin' Rose** *Top Fuel car. This fire made the ex-Top Gas pilot wish he was on the vast open spaces of the Bonneville Salt Flats. Gates' cookout took place at the Ontario Motor Speedway opener in the early 1970s. The car was extra crispy, but Gates escaped with minor burns. He returned to the salt flats after his asphalt adventure.*

A class act when he ran Junior Fuel and a class act when he ventured into Top Fuel was San Diego's Bob Williams. This photo was taken in Salt Lake City in the late 1960s. Unfortunately when the team headed home that evening, they were in a horrendous highway accident that claimed a crewman and almost killed Williams. Fighting back from his life-threatening injuries, Williams eventually recovered enough to race again.

Looking good, the restored Winkel, Trapp, and Hampshire Magicar in 1992 after it had been sold to Jim Lind in 1967 or 1968 and the NorCal racer ran it as the Uptight Out of Sight *AA/FD. It had disappeared for about twenty years and then wha-la there is was in its old glory.*

Nando Haase says "See ya!" to John Mulligan in the Ward and Wayre Longshot *Top Fuel car. Haase won Top Fuel Eliminator at this 1965* Hot Rod Magazine *meet.*

A cover for Drag Racing USA in the making; The Syndicate *twin Top Fuel car cruised through a Jack in the Box drive-thru with the late Phil Ditmars at the controls. Drag Racing USA loved the unusual covers I came up with in the early 1970s.*

The Leprechaun *Top Fuel car, seen here in 1965, never found the pot of gold at the end of the rainbow. Maybe a sponsorship from Lucky Charms cereal would have been in order.*

The last front-engine slingshot Top Fuel car to try to compete in Southern California was owned by Dan Horan, a low-buck racer with a lot of heart. Unfortunately, heart didn't win against those new mid-engine Top Fuel cars.

Qualifying at the 1967 NHRA Winternationals in Pomona was Gene Goldman driving the Crietz, Greer, and Goldman Top Fuel car (far side) versus Jerry Bivens in the Bivens and Fischer Checkerboard. *With tires blazing, Bivens pulled ahead, but not for long—that's a rod sticking out of the B&F Hemi. Needless to say, Goldman smoked on by and Bivens clicked it off.*

Coming out of our nation's capital was Bruce Wheeler and his Wheeler Dealer *Top Fuel car. Little Al Friedman did the driving for the beautiful candy red car. I only saw this car in 1968, then and I heard that Wheeler sold it and moved to Hawaii.*

A very popular team in Southern California racing circles was the **Groundshaker***, which consisted of a Top Fuel dragster and a ground-thumping AA/Fuel Altered. The Top Fuel car was "senior," the roadster was "junior," and the team was Bill and Mike Demerest, Glen Way, and Gary Read. Read drove senior and Way drove junior. Sometimes Read drove junior and Bill Demerest drove senior. Maybe Tom Ferraro drove junior? Confusing?*

The late Denny Milani blistered the Riverside, California, quarter-mile with the late "Terrible Tempered" Ted Gotelli/Milani Top Fuel car, running 7.55 at 210.76 mph in 1965. Then Milani faced Mike Sorokin in the Skinner, Jobe, and Sorokin Surfers car in the first round. Milani went "night night" and Sorokin was long gone. Ironically, the sport of drag racing lost both of these drivers in tragic crashes, Milani in August 1965 at Half Moon Bay and Sorokin on December 31, 1967, at OCIR.

Nineteen sixty-nine was supposed to be a banner year for The Fighting Irish team of Tim Beebe and John "Zookeeper" Mulligan. The Irish defeated Don "The Snake" Prudhomme for Top Fuel Honors at the NHRA Winternationals. However, at the Nationals, Mulligan raced Ivo in the first round, and had a fiery crash at about the 1,000-ft mark. He clung to life for two weeks before he passed away. Drag racing truly lost a favorite son.

With the Rader Wheels Special, *Nando Haase and John Smyser were Top Fuel winners at the* Hot Rod Magazine *race at Riverside Raceway in 1965.*

Prudhomme in the Hawaiian *heads for the finish line, and Billy Scott heads for the stars with the* Scotty's Muffler *Top Fuel car at Riverside in 1965.*

Pomona, 1967 NHRA Winternationals: Don Garlits with an all-red Top Fueler? Garlits soon realized his **Swamp Rat** *only won when it was painted black. He changed the color back to black and was able to win again.*

Ronnie Goodsell's **Earthquake** *Top Fuel car was a rolling experiment in the late 1960s with a different style of chassis and suspension. The "Quake" ran well, and its trademark smoking front wheel made it fun to watch and photograph. The car launched hard, carrying its left front tire. At about 300 feet the wheel dropped to the asphalt and smoked the tire.*

Fire burnouts were fun to photograph if they were planned, but if the VHT in the bleach box accidentally ignited, things could get very interesting, very fast. Norm Reis' crewman found out the hard way, with the VHT bottle aflame and his pant leg ablaze. I got the photos and he got hot. Luckily, he wasn't burned, but the Reis Top Fuel car had nice hot slicks. It's no surprise that NHRA disallowed fire burnouts many years ago.

The late, great Keith Black was a maker of horsepower, a lot of Hemi horsepower.

Sid Waterman was an innovator and Top Fuel racer who, along with longtime race partner Ronnie Hampshire, campaigned front-engine Top Fuel cars throughout the 1960s. Waterman's engine-building and fuel-system business expanded and it became a full-time job.

This dapper gentleman is Mike Doherty, the editor/publisher of Drag Racing USA *magazine. Mike turned* Drag Racing USA *into the number-one drag racing magazine in the country. If you had your car in* Drag Racing USA, *you had made it. Previously, Doherty was editor of* Drag World *and editor of the* Drag Racing Almanac *in 1966 and 1968.*

In 1957, Tommy "Watchdog" Allen broke into drag racing with a Chevy D/gasser. He held the record for D/G in Colton, California, at 98 mph. Here it's the 1960s and Allen is thundering in the Soapy Sales *Top Fuel car. Look out or the "Watchdog" will get ya!*

There is nothing in any form of racing like front-engine, nitro-burning Top Fuel at night.

One of the nicest Top Fuel cars Danny Ongais owned and drove was the Honda of Wilmington entry. This mid 1960s Top Fuel car didn't last very long. While qualifying at Irwindale, Ongais was on a banzai run when at about the 1,000-ft mark, the nitro-burning Hemi gave up its crankshaft. The crank came out and dug into the racetrack, causing the car to go up and over, nose first. Ongais was clocked at over 200 mph, upside down. When the smoke and mayhem died down, Danny emerged unhurt, but with a mangled Top Fuel car.

The faster Top Fuel cars went in the late 1960s, the harder they were to stop. Dual chutes made by Bill Simpson became popular, and here Dwight Hughes slows the Berry Brothers and Hughes Top Fuel car with its dual-chute system.

In the late 1960s John Bateman ran one of the most beautiful Top Fuel dragsters in Southern California. His **Atlas Oil Tool Special** *was a work of moving art. The swoopy Top Fuel car was not the first Bateman had run, but it was the prettiest. It ran as good as it looked with veteran Top Fuel driver Gerry Glenn at the wheel.*

In the mid and late 1960s Joe Winter ran quite a few Swinger *Top Fuel cars with drivers that included Harry Payne, Ron Goodsell, Frank Pedregon, and Kenny Logan. Of course, Winter drove as well until he lost his leg to cancer and Logan began his stint in the* Swinger. *Winter came back to run his* Swinger *Funny Car until cancer eventually claimed his life.*

The Atchison, Topeka, and the Chevrolets: John Peters' awesome Freight Train *twin Chevy Top Fuel gas digger was awesome. Whether it was driven by Goob Tuller, Floyd Lippencott Jr., Wild Bill Alexander, or Billy Scott, the* Freight Train *rung up many a victory over the years, not to mention winning the first Professional Dragsters Association race. Now, the question is do you know who Floyd Lippencott Jr. was?*

OCIR always had a bite, and here Ted Inyo and his San Diego-based Top Fuel car discovered that. Ted did three of these on this run in the early 1970s and bent his race car.

In 1962, Don Moody ran a 7.76 with the Davis and Moody digger at Pomona. After this Moody had a very bad crash at Lions and seemed to retire to his job at Engle Cams. Fast forward to the late 1960s, and he was back driving the Cerny Lins Top Fuel car. With partner Wes Cerny, Mood became a big gun in SoCal Top Fuel racing.

Rich Welch does his best Sergeant Schultz impression. Welch was a local photographer who sold me my first 35-mm camera. I also traveled to Los Angeles with Rich and his brother Les in 1965 to the Hot Rod Magazine race at Riverside. Unfortunately, Welch was colorblind which severely hampered his drag racing photographic career, so he gave up photographing the race cars and opened a speed shop to sell parts to the racers in the mid-1960's.

A tale of woe for Fred Farndon at Pomona in the late 1960s: It was a very cold weekend at the Winternationals. The Top Fuel cars were having problems starting, so they were pushed from the upper build-up lanes. Once fired, the cars had to turn left to line up to run down the quarter-mile. It was getting late in the day and the fuel cars had been sitting for quite awhile when they finally gave the signal to push out. Away they went, faster and faster, down the fire-up lane. There was no fire in Farndon's fuel-burning Hemi and they were getting closer and closer to the hard left turn. Finally, there was life in the Hemi but Farndon was going too fast to make the turn. Meanwhile, a spectator was going into the pay phone booth that rests about 50 feet from the apex of the turn. If Farndon turned now, he'd roll the car, so he went straight into a steel pole and fence, tearing off the front end of the car, but the rest of car kept going right up to the doors of the phone booth with a very surprised spectator inside. Farndon clicked off and errant car was pulled away from the phone booth. Nobody was hurt, but the car was a lot shorter.

SLINGSHOTS FROM AROUND THE COUNTRY

The Midwest had some of the best of the best Top Gas dragsters in the country. Butch Hummels' twin Top Gas dragster from Ohio blistered many a Midwest quarter-mile.

Relocating to Los Angeles meant getting to know more racers in Southern California and from other states as well. One of my first ventures out of California was to the NHRA Spring Nationals at Dallas International Raceway when longtime Texas photographer D.D. Cleveland invited me to the Lone Star State. At the racetrack he introduced me to many Texas Top Fuel racers, including Vance Hunt—a good guy with a black hat—and his driver, Watus Simpson. Another Texas racer was George Cox, owner of the *Banzai* Top Fuel car, driven by Fred Rowsey. Richard Tharp and Gary Bailey were couple of wild men—these guys could drive anything that burned nitro and had four wheels.

When discussing Texas Top Fuel racers, the Carroll Brothers and Oxman are at the top of the list, along with their driver, the smiling Buddy Cortines. Also significant were the "Chaos Kids" (the Tartar Brothers with Mike Tartar at the controls), Raymond Austin and Joe Haggerty, and the Simpson Brothers. Dan Rightsell, Bob Holley, Marshall Love, Jackie Peeples with Chip Woodall driving, and Jimmy Duppstadt with Bob Gibson driving also raced in the Texas Top Fuel wars in the late 1960s and early 1970s. Hardcore Top Fuel fans are probably saying, how come you didn't mention Bobby Langley or Eddie Hill? Langley had retired in the mid-1960s and Hill quit Top Fuel racing for a time to go drag-boat racing.

Just north of Texas is Oklahoma, and this northern neighbor produced some very fast Top Fuel racers, including Jimmy Nix, who flexed his Top Fuel muscle by

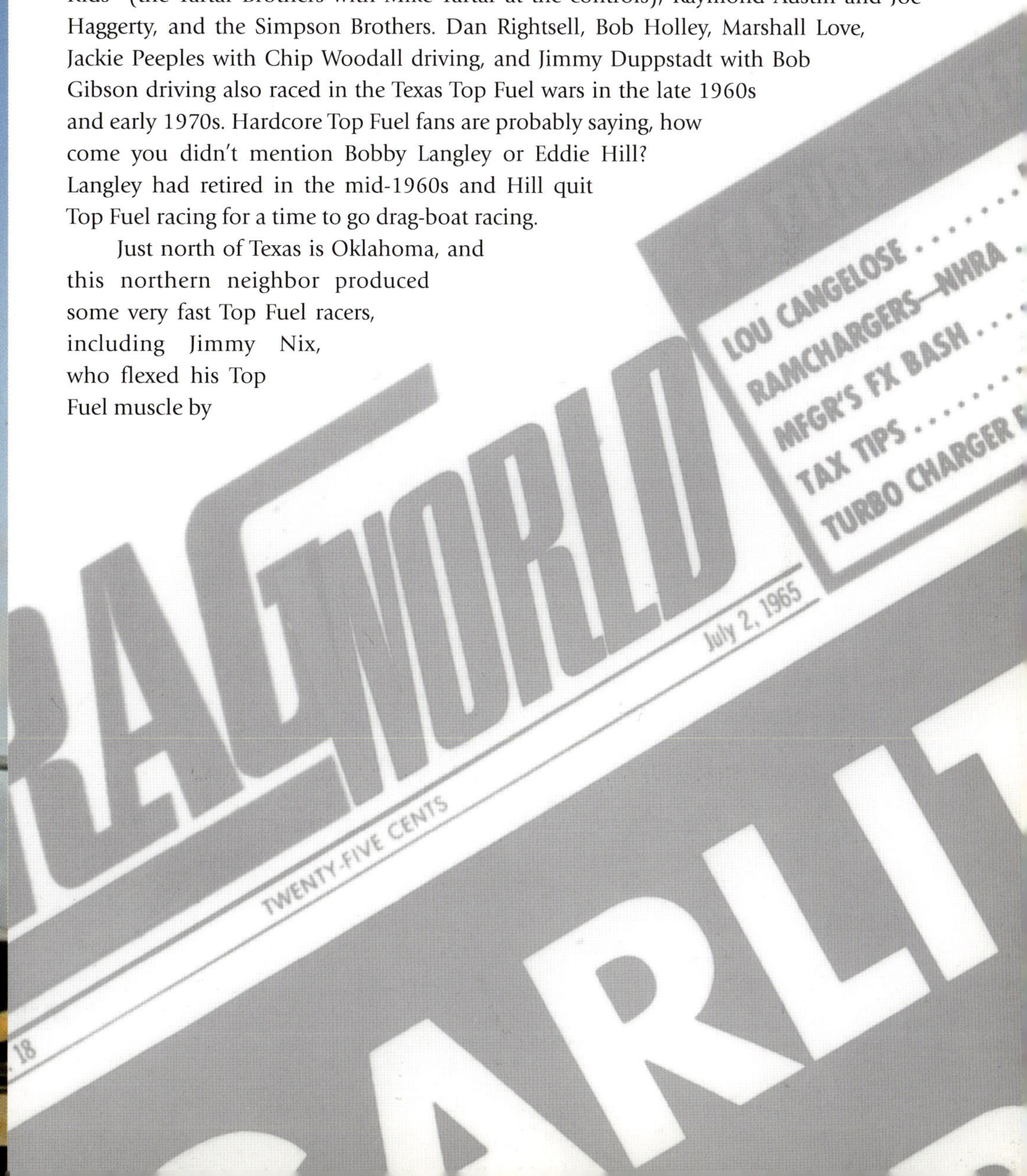

"Flamin' Frank" Pedregon began his racing career in 1958 in Texas with an A/C coupe. Five years later in SoCal, he built his first nitro-burning Hemi engine with a budget of $35. He debuted his flaming, tire-burning coupe on February 20, 1965, at Lions Drag Strip. Soon the legend took root and Pedregon became known as someone who could build something from nothing. He took junk parts from other race teams and ran with them. After racing and crashing his famous flaming coupe, Frank raced in the Top Fuel class for the latter part of 1966 and into 1967. Pedregon claimed to be the "World's Fastest Mexican" while NorCal racer Frank Martinez also claimed the title. As far as I know, it was never determined which one actually had the title because they never met on the battlefield of asphalt.

winning the 1966 NHRA Springnationals. "The Wizard of Oz" Bennie Osborn took the 1967 NHRA World Finals Top Fuel title, and of course, "The Mad Bomber of Tulsa" Bob Crietz and his partners Greer and Goleman were a feared trio in Top Fuel. Okie transplants Steve Carbone (Crietz, Greer & Goleman), Connie Swingle (Don Garlits), and Gary Ormsby (*The Vagabond, Cheetah*, and his own Top Fuel car) made their presence known in Florida, NorCal, and SoCal Top Fuel racing during this time. The Okies were big-time contenders in Top Fuel during the late 1960s.

To the east of Texas is Louisiana, home to Paul Candies, who was partnered with "Q-Ball" Wale for the Candies and Wale Top Fuel car during the mid-1960s. On July 17, 1965, Q-Ball was killed in a crash at Monroe, Louisiana. Robert Anderson, with drivers Kerry Macinroth and Ronnie Martin, was another who made Louisiana proud in the late 1960s and early 1970s. In 1970, Dave Chenevert was named one of the toughest in the Southeast to race in Top Fuel, and he won the first-ever NHRA Gatornationals driving Anderson's Los Angeles-based Top Fuel car. Soon after that win, Chenevert returned to his Shreve Automotive Top Fuel car. Last but not least was Top Fuel racer George Donaldson.

These Southeastern Top Fuel guys were a great group of racers, and they always had a smile and a handshake for me. But the Californians were the hardest-racing group. That is my opinion and I am sticking to it.

When you break your Top Fuel car, don't cry about it, just drive a friend's No.2 car. Gary Ormsby did just that at a race in the late 1960s when Jim Brandy needed a driver for his second Junior Fueler and Ormsby needed a ride for a Saturday race. The veteran driver showed the fans he loved to drive, no matter what kind of digger.

As the "Wizard of Oz," Southwestern racer Bennie Osborn was the king of the Top Fuel hill in 1967 when he became the 1967 NHRA World Champion in Top Fuel. A big win like that and your racecar suddenly has a full body and cool paint job.

It's that team of Okie and California guys with the "Mad Bomber" of Tulsa, Bob Creitz, teamed with Ed Donovan and Butch Maas at the NHRA Nationals in the early 1970s.

The great Northwest was home to one of the quickest blown fuel-burning Chevy diggers in the late 1960s. It belonged to John Garrison, was driven by Jerry "The King" Ruth, and could run with the blown fuel Hemis. Ruth drove Garrison's King Rat *when he parked his Top Fuel car and wanted to break somebody else's parts. Like all the blown Top Fuel Chevys, it soon faded into drag-racing history.*

A very hard-running Southwest-based Top Fuel car belonged to the team of Graham and Groom with Marvin "The World's Fastest T.V. Repairman" Graham at the controls. Because they either ran great or blew up, I nicknamed the duo "Bam and Boom."

After Chuck Poole destroyed the M&R Top Fuel car at Vacaville in 1965, the M&R duo came back with an all-new entry driven by Russ Lewis. The trio never did gel into a race team, and Lewis ended up crashing this car and was burned.

A great example of a late 1960s Top Fuel car was Harry and Maxine Lehman's American Way digger. The late Al Friedman was the driver of this potent East Coast Top Fuel car.

Michigan has produced some great drag-race owners/drivers in the nitro classes. One of those drivers was Chuck Kurzawa, who was the last driver of the Ramchargers Top Fuel car before they switched to Funny Cars. Kurzawa found a home driving for the Bob's Drag Chutes Top Fuel car.

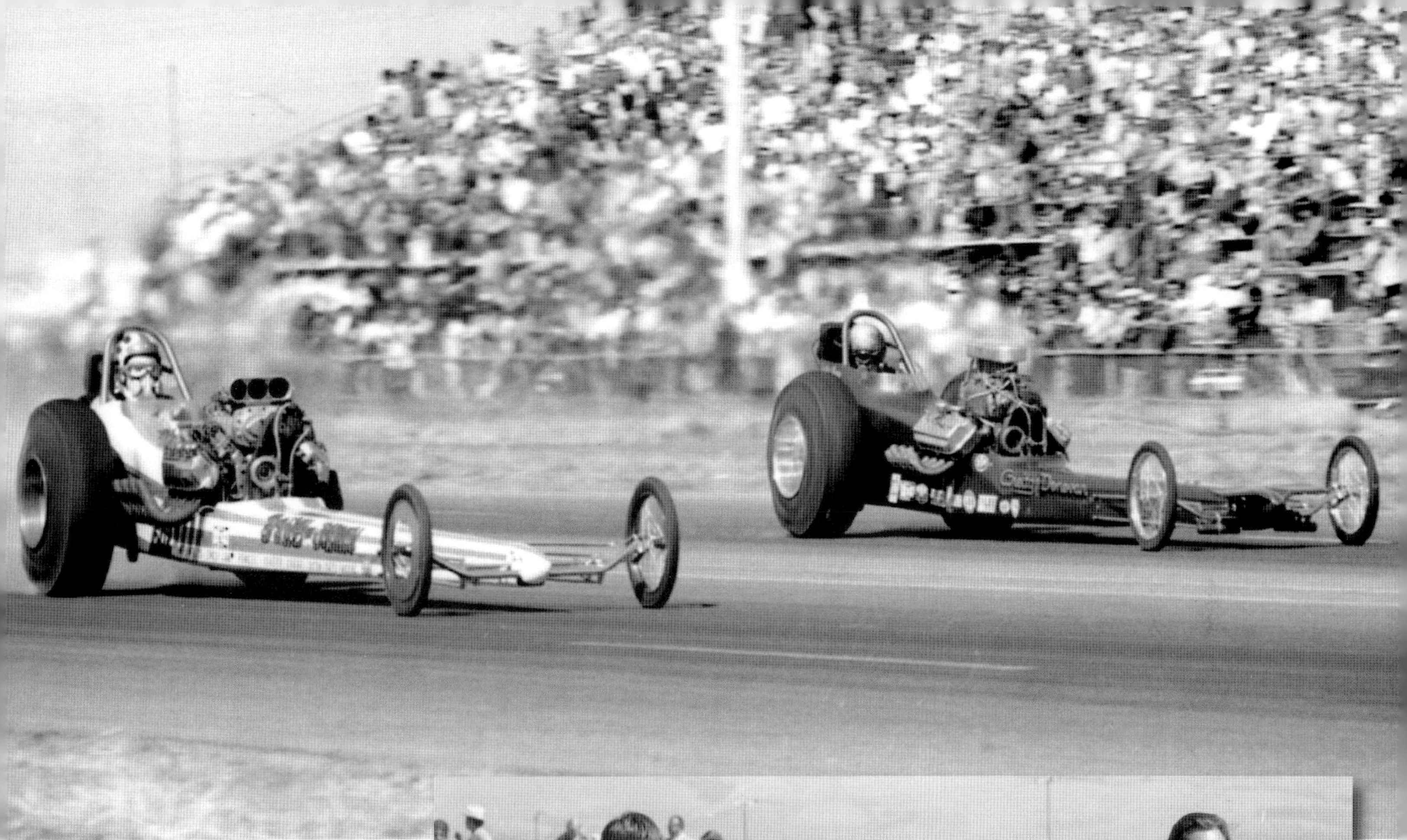

Neck and neck down the quarter-mile was Dailey and Johnson (near side) versus Creitz, Greer, and Donovan. The CG&D Top Fuel car was tuned by "The Mad Bomber" of Oklahoma, Bob Creitz, who earned the nickname by dropping ball bearings on a major city in Oklahoma from a small private plane. It was just a joke, but the police didn't find it very funny and the local media named Creitz "The Mad Bomber."

If your 426 Hemi is breathing death, just tape some rags around the breathers and run. That's what the Bob's Drag Chutes Top Fuel car did in the early 1970s.

Comin' out of corn country was John Dearmore's beautiful Top Fuel dragster. Dearmore was from Kansas and his drivers were Byren Budd or Norm Gingrass. His full-body 392 Hemi-powered dragster raced all over the Midwest in the late 1960s. Dearmore did well match racing and running the NHRA World Champion Series circuit.

Vance Hunt was a legend in Texas Top Fuel racing. Hunt is the one in the black cowboy hat holding on while driver Watus Simpson readies to fire the Hunt and Simpson Top Fuel car on the rollers at the old Dallas International Raceway in 1969.

Heading for the photographer on the ladder is Metairie, Louisiana, Top Fuel hero Dave Chenevert. Chenevert's ride is the Shreve Automotive Top Fuel car. Dave drove Robert Anderson's Top Fuel car to win the first Gatornationals in 1971.

The Glenn Company started shooting the full-length feature film Drag Racer *in June 1972, and I was there shooting the stills and magazine coverage. They had a great cast of cars: the* Freight Train*, Snively and Annin, Gerry Glenn, Dennis Giesler's AA/FA, Sherm Gunn's AA/FA, and a few more.*

Rolling Thunder *from Texas was the Carroll Brothers' awesome twin-engine nitro burning Top Fuel car. Veteran driver Buddy Cortines drove the* **Texas Whips** *twin. This car shook the pavement when it ran. I didn't really care what this car ran ET or speed-wise, I just loved seeing and feeling it run in 1970.*

This odd Top Fuel car was built by Jim Hume and placed the driver almost in an upright position. The driver could see great—right into the guardrail he was going to hit. The rear end had suspension and was evil handling, but fun to watch in 1971.

One of the most beautiful Top Fuel dragsters ever to traverse the quarter-mile came from the great Pacific Northwest. I first saw the Eagle Electric Top Fuel car at Pomona in 1967. This sleek full-bodied racer ran as good as it looked and churned the tires with Frank Rupert at the wheel.

DRAG WORLD

VOL. 4 NO. 26 — FRIDAY, OCTOBER 4, 1968

FEATURE INDEX

Prudhomme-Ramsey	2
Keaton Top Funny	4
Roberts' Hole Shot	6
Smoker Battle	11
Carelessness	16

CREITZ-GREER-DONOVAN & CARBONE WIN.....

CARBONE TAKES $3,000 AT FREMONT PDA MEET

Steve Carbone, driving the Creitz, Greer, and Donovan Top Fuel car from Oklahoma, won the 1968 Professional Dragsters Association race at Fremont. Carbone waded through a stellar field of Top Fuel dragsters and defeated Warren, Coburn, and Miller in the final.

Can you guess whose former Top Fuel car this was? Bill Butler of Reno, Nevada, purchased and drove the famed Greer, Black, and Prudhomme Top Fuel car. The former GBP racer was repainted and named Bonanza. *The car was outdated and had a problem with wheelstands. I didn't mind, but I know Butler wasn't too thrilled.*

The last hurrah for the front-engine Top Fuel dragsters was in 1971. The Jade Grenade *was state-of-the-art at the time, but soon all front-engine Top Fuel cars were obsolete. These are photos of the original* Jade Grenade *shot at Old Bridge Township Raceway Park in Englishtown, New Jersey, in 1971.*

Mitchell, Hay, and Kleintz's Top Fuel car out of the Midwest did the wing thing trying to keep up with the onslaught of rear-engine Top Fuel cars. It didn't make much difference, but at least driver Jack Mitchell had shade.

We're at the eighth-mile at the 1970 NHRA Nationals and the Gopher State Timing Association Top Gasser is doing its impression of the Little Red Wagon.

Two Top Fuel cars in one lane is not a good combination. Bennie Osborn (right) is trying to give the Gary Baily Top Fuel car a lot of room so there won't be a meeting of the diggers at Dallas in 1970.

In the early 1970s a northwest Top Fuel team started to make its mark in the nitro class, and that team was the terror of Herm Petersen and Sam Fitz. With Herm "M-80" Peterson at the helm, the team was a threat at any race it attended. Peterson has the height record for 55-gallon drums. Armed with M-80s, he sent a drum sky high in the pits after the races. I helped by going through Missouri and picking up cases of M-80s for him.

Dale Thierer and the awesome Hemi Hunter nitro-burning Top Fuel Chevy were thunder on the East Coast. Thierer was deadly in the early 1970s.

Bob Gibson has his own personal barbecue with Ray Godman's **Tennessee BoWeevil** Top Fuel car. The starting-line personnel at Bristol, Tennessee, quickly put out the flames.

In the early 1970s, a long, mean, green machine was stalking the East Coast Top Fuel class with the name of Jade Grenade. *The team of Thomas, Lenhoff, and Furher teamed went match racing all over the eastern United States with Teddy Thomas guiding the green machine down the dragstrip.*

The "Cincinnati Kid," Jim Bucher, ran a twin blown Chevy in Top Gas during the late 1960s. Bucher had one of the first toy sponsors in drag racing. Kenner Toys of Ohio even made toys of Bucher's race-car in their Kenner SSP toy series. Bucher soon vacated Top Gas for Top Fuel.

DRAG WORLD

FEATURE INDEX

VOLUME 2, NO. 4 TWENTY FIVE CENTS JANUARY 28, 1966

Year Of Snake Over

GARLITS GETS PRUDHOMME

Love him or hate him, Don Garlits is the king of dragsters. The Florida "Swamp Rat" had been kicking racers' butts for a long time in Top Fuel. His favorite racer to beat was Don "The Snake" Prudhomme. Garlits felt Prudhomme was his No.1 nemesis in Top Fuel.

The late 1960s saw Jim Paoli emerge onto the Top Fuel scene. The Midwest-based Paoli brought his Yankee Pack Rat to major NHRA events and made his mark in the Top Fuel class.

It seems the only Northwest Top Fuel driver to ever get any press was Jerry "The King" Ruth, but making its mark in the Top Fuel class in the Northwest and beyond was the team of Daily and Johnson. In the late 1960s this duo impressed many a racer and fan.

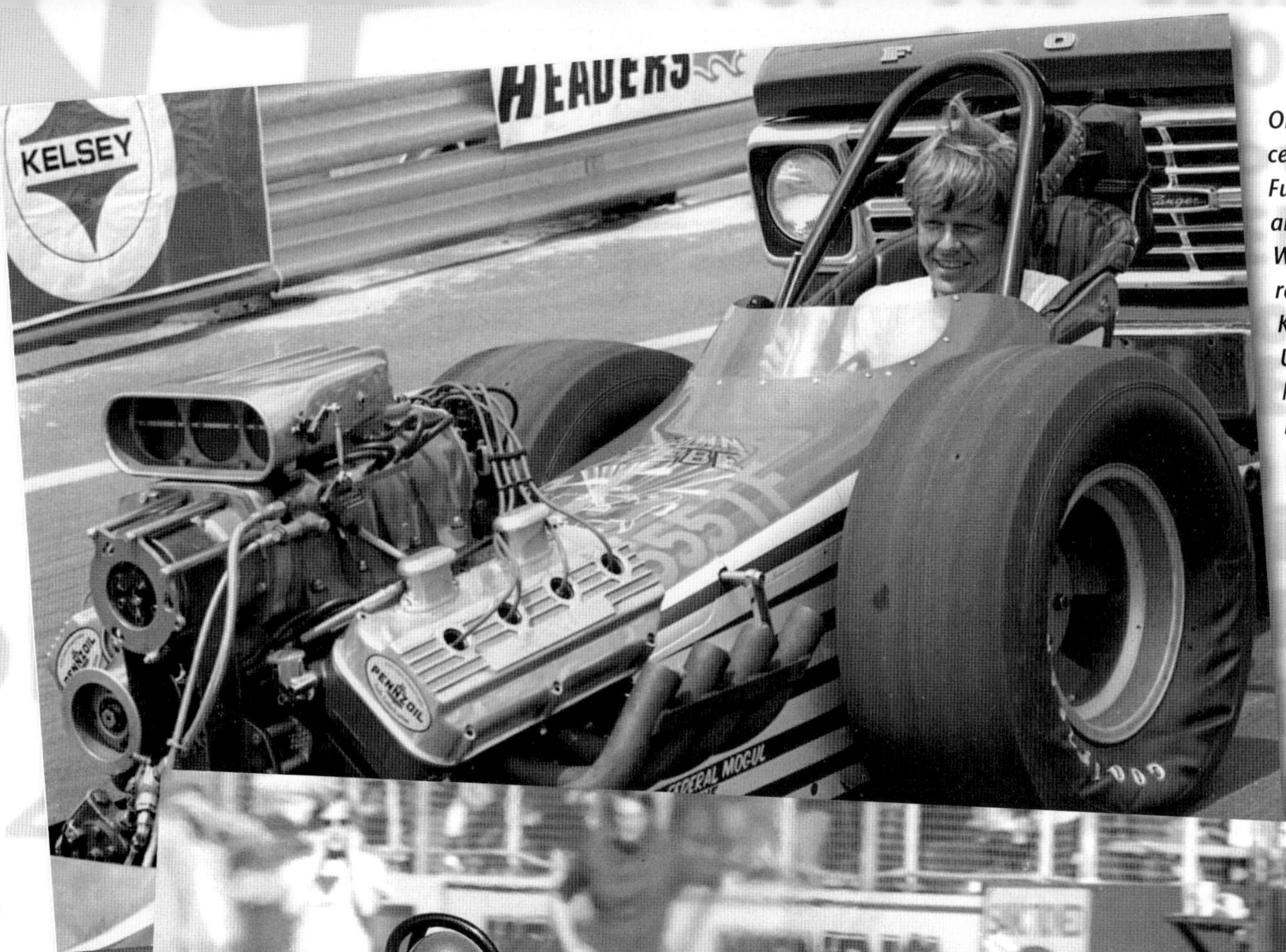

One of the most successful independent Top Fuel racers in the 1960s and 1970s was John Wiebe. The smiling racer from Newton, Kansas, traversed the USA, competing in NHRA and AHRA national events and match races. He went head to head with all the big-name Top Fuel teams of the time and won his fair share. When Wiebe was in SoCal to race, he stayed at my place.

REYES' TOP 32 OF 1969-71

In 1969, the sport of organized drag racing was 21 years old and getting bigger and better every day. Even with the invasion of the Funny Car class, Top Fuel was going strong. The backbone of the class was a network of hardworking hot rodders who loved their sport and loved speed. Most ran a 426 or 392 Hemi on nitro and with the development of slipper clutches they were even going faster than with nitro alone. As they went faster the safety issues became greater as they had more fires and blower explosions. Most Top Fuel drivers who lived to talk about this time period have their fair share of burn scars to tell their story of how dangerous it was to drive a front-engine dragster.

I chose 1969-71 because it was the last couple of years that many of these heroes campaigned in Top Fuel. The draw of more bookings and money soon lured them to the Funny Car class. This section is all about those Top Fuel racers who, in 1969, were at the top of the class and would venture into Funny Car.

1. "Big Daddy" Don Garlits
Garlits, from Seffner, Florida, had been the one to beat for 15 years (1954 to 1969) in his never-ending versions of the **Swamp Rat**, *and that continued into the next decade. During 1969, he ran both the* **Swamp Rat XII** *and* **XIII**, *both with 426-ci Dodge Hemi engines. The* **XII** *had a full body while* **XIII** *was only half bodied. Here Garlits makes a banzai pass at Indy with his newly restored* **Swamp Rat I**, *which was No. 1 in 1984. This car featured many innovations when it debuted in the 1950s. Garlits won a lot of races with this car.*

2 & 3. Don "The Snake" Prudhomme & Tom "The Mongoose" McEwen

Prudhomme was from Granada Hills, California, and in 1969 was behind the wheel of his Wynn's Winder, *which was a half-bodied dragster with a 426 Chrysler Hemi in it until he put a full body on it toward the end of the season. He also won the U.S. Nationals in 1969. Tom McEwen was from Garden Grove, California, and he drove either his full-bodied Top Fuel car or a Funny Car often in match races around Southern California. In 1971 "Snake" and "Mongoose" became the "Wildlife" racing team sponsored by Mattel's Hot Wheels. Top Fuel and Funny Cars kept Prudhomme and McEwen busy well into the 1970s.*

4. Connie Kalitta
Warming up in 1966 at Irwindale was Connie Kalitta and his SOHC Bounty Hunter *Top Fuel car. You might say Kalitta got hot less than five weeks later in 1967. He beat Gene Goldman to claim Top Fuel at the NHRA Winternationals in Pomona.*
Kalitta, out of Mt. Clemens, Michigan, was a longtime Top Fuel racer who switched to Funny Cars. He drove Pancho Rendon's Top Fuel car at Indy and managed a great crash. In 1971 he brought a previously unknown woman, Shirley Muldowney, into the nitro classes and together they campaigned the Bounty Hunter *and* Huntress *Funny cars (see my other CarTech book* Funny Car Fever*).*

5. "T.V. Tommy" Ivo
In April 1967 at Fremont Tommy Ivo and Connie Kalitta went head to head for the No. 3 spot on the Drag News *Top Fuel top 10 list. Ivo prevailed two straight and sent the SOHC Ford Fueler home to Michigan empty-handed. Based in Burbank, California, Ivo was the touring king of Top Fuel with 72 dates a year. He drove a Don Long-built Top Fuel car with his name on it and came with a complete extra spare car. When he toured, the trailer had a large glass window for the public to see the cars as he traveled down the highway. Ivo spent the first half of the 1970s racing Top Fuel and then switched to Funny Cars. Ivo's favorite food on tour was Howard Johnson's fried clams and he also collected motel room keys, which filled a dresser drawer at home.*

6. Chris "The Greek" Karamesines
There is only one crazy Greek in Top Fuel racing, and that is the legendary Chris Karamesines, one of my early heroes in Top Fuel. At 77 years old he still races IHRA and a few NHRA races each year. He believes in staying with a good thing as all his cars have been named the Chizler. *The Chicago racer later built a Mustang Funny Car, drove it a couple of times, sold it, and went back to his Top Fuel car, then built another 'Cuda Funny Car driven by his buddy Cliff Zink.*

7. John "The Zookeeper" Mulligan
Based in Garden Grove, California, Mulligan won the 1969 NHRA Winternationals in the Beebe and Mulligan Fighting Irish *Top Fuel car. Later in 1969, Mulligan succumbed to injuries received in a fiery crash at the 1969 NHRA Nationals. Here he is in the Ward and Warye* Longshot *against Dave MacKenzie in the Herbert and Reath Top Fuel car. Mulligan took this race.*

8. Jim "Superman" Nicoll
When "Superman" lit up the Dallas International Raceway bleach box, the question was, accidental or intentional? Nicoll will never tell, but he got his picture taken and his mission was accomplished in 1970.
From Spring Valley, California, Nicoll match-raced and toured the United States with his Top Fuel car Der Wienerschnitzel. *He created a memorable final at the 1970 NHRA Nationals when he raced Don Prudhomme. A clutch failure blew his car in half at the finish line, and the spectacular crash was recorded by ABC Wide World of Sports. Nicoll walked away from the crash and soon switched to Funny Cars.*

9. Don "Mad Dog" Cook
Don "Mad Dog" Cook shows us why the caps on fuel tanks were changed from snap down to screw on. A nitro bath isn't too refreshing at 200 mph. He was okay but a little wet.
Hailing from Walnut Creek, California, Cook stayed with Top Fuel, driving a car that was painted gray with a few decals on it. As he became a star on the AHRA circuit, he found a little help to upgrade his car and eventually named it Southwind and when he switched to a rear-engine car during the 1970s he named that car Southwind II.

10. Jerry "the King" Ruth

"King of the Northwest" was the very humble title Jerry Ruth claimed for himself. Ruth was a true drag racer. His reign in Top Fuel reached all over the USA. He started at age 16 with a 1937 Chevy Gas Coupe and the rest was history. In this photo, he drives around Ken Safford in front of a standing-room-only crowd at Lions. A little-known fact about Ruth is that he had one of the first two-car teams in drag racing that included a Top Fuel car and a Funny Car, both of which he drove. He stayed with his blue-and-white-striped Top Fuel car with a 392-ci Hemi engine and had Frank Hall and Lee Beard at the controls of the Funny Car.

11. John Keeling and Jerry Clayton

The most beautiful slingshot dragster ever to grace any quarter-mile was the California Charger. Owners Keeling and Clayton did it all first class. The great paint and chrome didn't win races, but a great team and driver Rick Ramsey did. The team's biggest win was Top Fuel Eliminator at the first NHRA Supernationals in 1970.

12. "Sneaky" Pete Robinson

In the 1960s, the southern gentleman of Top Fuel racing was the late "Sneaky Pete" Robinson. This soft-spoken Georgian was an innovator, engineer, and longtime drag racer, both in Gas and Top Fuel. Robinson won Top Eliminator at the NHRA Nationals in 1961 and was the NHRA Top Fuel World Champion in 1966. He scored the first Top Fuel win for the Ford SOHC motor in 1965. Robinson lost his life at the 1971 NHRA Winternationals in Pamona. He had equipped his car with a downforce innovation he called "the vacuum cleaner." It worked too well, and on one run it created so much downforce that the front tires of the Top Fuel car exploded and Robinson hit the Pomona guardrail at speed right in front of me. Drag racing lost a great person on that Saturday.

13. Tom Hoover

Tom Hoover made his first appearance at Fremont a memorable one when on this pass he did a wheelstand and ran off the end of the track. He missed me by about 6 ft. Then he came back and qualified. After the race, Hoover's rig was stolen and nothing was ever recovered. This Minnesotan also became a class act on the nationwide Funny Car circuit with his White Bear Dodge Funny Car. He won a lot of local Top Fuel races as well as a few AHRA races. In 1965, he won the AHRA Top Fuel Championship and in 1968 he won the Olympics of Drag Racing.

14. Leroy Goldstein

Leroy Goldstein gave the local fuel racers driving lessons from NorCal to SoCal. The first time I saw the team race was at my home track in Fremont, and the B&G team swept through the Top Fuel field. On the final, Goldstein entered the finish-line lights ahead of his NorCal final-round challenger. Reaching for the chute, Goldstein noticed his steering wheel was not attached to anything. No panic, the chute came out and the car stayed straight and stopped on the racetrack. Goldstein, from San Diego, turned over the Ramchargers Top Fuel car to Chuck Kurzawa and went fuel Funny Car racing. Goldstein became one of the best and most feared Funny Car drivers ever.

15. The Ridge Route Terrors
No strangers to the winner's circle at SoCal drag strips, Warren, Coburn, and Miller hit it big in 1968. The Bakersfield-based trio marched through a stellar field of Top Fuel dragsters to claim the Top Fuel title at the 1968 NHRA Winternationals. With James Warren at the wheel The Ridge Route Terrors kept on racing and winning in Top Fuel until all of the SoCal racetracks were gone.

16. John Wiebe
At old Dallas International Raceway in 1969, "Kansas John" Wiebe was getting ready to launch his nitro burner into the Texas night. Wiebe was a hardcore racer from Kansas and he and his loyal crewman Steve Phillips raced AHRA events and match-raced all over the USA. He ran a few major NHRA events when they didn't conflict with his AHRA and match-race events. Wiebe was one of the last of the front-engine dragster owner/drivers to switch to the rear-engine Top Fuel cars in the late 1970s because he was a bucks-down racer who had a limited budget and often had to do everything himself. When it wasn't fun anymore, "Kansas John" retired.

17. "Little" Dave Babler
Based in Santa Monica, California, Babler was one of the first touring professional Top Fuel racers. His California Woody, *named for its wood inlay on the body and rails, was runner-up to Jim Dunn at the 1969 Bakersfield Smokers Meet. After that, the Top Fuel owner/driver left the tour and was never seen at a drag strip again. The legend is that Babler found himself a Minnesota milkmaid and settled in the Midwest.*

18. "Big Jim" Dunn
Before Funny Car Summer, Jim Dunn, from La Mirada, California, was a Top Fuel stud. The La Mirada fireman was a fixture in nitro racing in Southern California in the mid to late 1960s. Like many a Top Fuel racer, Dunn worked his way up to a fuel digger by driving a blown Fiat, a Top Fuel Chevy, a Top Fuel Hemi, and then a Woody Gilmore-built Funny 'Cuda. He ran one Top Fuel car with Al Yates, and it was a powerful force to be reckoned with. The Dunn and Reath Top Fuel car, a partnership with Joe Reath, was a consistent visitor to the winner's circle at California dragstrips. He then took the rear-engine concept to Funny Car racing and made it work.

19. Dennis Baca

Out of Walnut Creek, California, Dennis Baca just kept getting faster and more professional in the late 1960s. His Jim Davis car now had a full body, and a big bad Keith Black elephant Hemi replaced his 392-ci Hemi. He made his mark in NorCal Top Fuel racing and in 1969 was ready to tour. Baca continued to race Top Fuel throughout the 1970s.

20. Carl Olson

It's the Grand Prix of drag racing at Irwindale in 1969. Carl Olson prepares to navigate Jack Ewell's Top Fuel car into the Southern California darkness. Based in Torrance, California, Olson thundered in the Ewell, Bell, and Olson Top Fuel car. Olson raced his own Top Fuel car at the beginning of the 1970s and then hooked up with Mike Kuhl to go Top Fuel racing. Kuhl and Olson won the 1972 NHRA Winternationals and toured nationwide, also winning IHRA and AHRA events.

21. Ron Rivero

At the 1968 NHRA Winternationals, two heavyweights in Top Fuel squared off in qualifying. The Ramchargers (far side) took on the K&G Speed Associates Frantic Four *from California/Pennsylvania. Neither won the event, but Ron Rivero and the* Frantic Four *claimed the Top Fuel title at the Bakersfield Smokers race about a month later. Rivero was from Pomona, California, but the team moved to Pennsylvania and brought about the* Frantic Ford *Funny Car.*

22. Steve "The Mandrill" Carbone
The SoCal journeyman Top Fuel driver from Downey, California, drove the Atlas Oil Tool Special for John Bateman. Carbone won a lot in that car, and then got a ride in the Crietz and Donovan Top Fuel car. He became a factor in any race he entered. The 1970 season was almost over when Carbone debuted his all-new Top Fuel dragster at the Ontario Motor Speedway Grand Opening ceremonies. Carbone's was the last front-engine car to win the NHRA U.S. Nationals, defeating Don Garlits in 1971.

23. Marvin Schwartz
A tennis pro and Top Fuel owner/driver, the late Marvin Schwartz was an early touring Top Fuel racer and good guy. He was one of the few to ever drive for Don Garlits. From Largo, Florida, the ex-Garlits team driver raced his Anaconda *Top Fuel car all over the USA.*

24. Bennie "The Wizard of Oz" Osborn

First round at the Bakersfield Smokers Event in 1969: Dwight Salisbury in The Beach Boys *Top Fuel car versus Bennie Osborn. Salisbury went boom while Osborn looked on. Nobody was hurt in the big boom and Osborn won the race. The former 1967 NHRA World Champ from Sand Springs, Oklahoma, continued to race off and on in Top Fuel, but the cost of racing began to get exorbitant, eventually taking him out of the game.*

25. Jimmy King and Donnie Marshall

From the smallest state in the USA, Rhode Island, came one of the East Coast's strongest Top Fuel teams—Jimmy King and Don Marshall. King drove and Marshall prepared the car. The duo scored Top Fuel wins at two major AHRA races in 1970. These guys and their El Diablo *were a threat wherever they raced. In the 1970s, the only Rhode Island Top Fuel team became the second Rhode Island fuel Funny Car team.*

26. Dick LaHaie

LaHaie, from Lansing, Michigan, was a very talented tuner, driver, and owner primarily in local Top Fuel racing. He ran well at AHRA and NHRA events getting as far as runner-up, but he didn't win the big prize until he won the NHRA World Championship in 1987. His daughter Kim got her Top Fuel license, but didn't pursue a driving career. She sure can tune a nitro-burning engine and make it go very fast.

27. "Surfer Hank" Westmoreland

This very talented driver from Malibu, California, shoed Jim and Allison Lee's Great Expectations*, a Virginia-based fueler, as well as and Don "The Beachcomber" Johnson's SoCal-based Top Fuel car. Allison Lee was the first woman crew chief of a Top Fuel dragster and could make some awesome horsepower. Eventually, Tom Raley got the* Great Expectations *ride and Don Johnson started driving his own Top Fuel car, so I guess Westmoreland went surfing.*

28. Kelly "The Controller" Brown
Brown, from Northridge, was one of the best journeyman drivers in Southern California. Leland Kolb and Lou Baney are just a couple who had Brown at the controls. Brown and the late Lou Baney teamed in the late 1960s and early 1970s to race the Folger Ford SOHC Top Fuel digger. Brown then jumped into Funny Car racing and did well in the Southern California area.

29. Kenny Safford
Veteran Southern California Top Fuel driver Kenny Safford came north to drive Ted Gotelli's Top Fuel dragster. Kenny replaced Roy Thode, who had replaced the late Denny Milani. The tall, lanky Safford, who was the original Sour Sisters *and* Shark *car driver, finished his Top Fuel driving career in Ted Gotelli's and Larry Bowers' Top Fuel cars. You need a scorecard to keep up with the musical drivers. In 1971, at the NHRA Winternationals, Safford drove Bowers' new Top Fuel car and scored a runner-up in Top Fuel to Garlits' rear-engine racer. Safford soon tired of Top Fuel and chose to drive Funny Cars for Stone, Woods, & Cook and Mr. Norm instead.*

30. Gary Cochran
Cochran was a Top Gas racer turned Top Fuel racer and one strong-running independent from Fountain Valley, California. He did very well as an independent Top Fuel owner/driver, and when he parked his digger, he drove for various other Top Fuel teams, including Carl Casper's Young American, *racing throughout the country at mainly NHRA national events. Cochran was a talented driver and a great guy. Funny Cars were also in his future.*

31. Larry Dixon
The Southern Californian in the Mr. Fireside Inn *roadster owned and drove the last front-engine Top Fuel car to win the Winternationals, defeating Tony Nancy in the 1970 final. Dixon then raced his own car and the thundering Howard Cams Dixon and Dunn Top Fuel Chevy. He was also one of the first eight drivers in the Cragar Five-Second Club for Top Fuel cars. This photo is from late in 1971 at OCIR. Larry Dixon was haulin' in the Dixon Family Top Fuel car. Check out the front wheels at 200 mph.*

32. Tony "The Loner" Nancy

Last but by no means least is Tony Nancy, from Sherman Oaks, California. I think Nancy's 22 Junior race cars had been around forever. From roadster to Top Gas digger, from rear-engine Top Gas digger to his fire-breathing **Wynn's Sizzler** *Top Fuel car, every one was a classic and done first-class. Nancy was one of kind, a showman and a gentleman racer. He put Wynn's* **Sizzler** *into the 1970 Bakersfield Top Fuel winners circle.*

So there you have my personal Top 32 of 1969-71. I know I left out a few names and cars, but I feel these 32 are the backbone of front-engine Top Fuel era of drag racing. No matter how many Hot Rod Reunions are held and no matter how many retro Top Fuel cars are built, the front-engine Top Fuel cars of drag racing's early days cannot be brought back. The cars are gone, most of the racers are gone, and that feeling, smell, and atmosphere at the drag strip have disappeared. I am very lucky to have experienced the true days of drag racing when its daring young men and women churned their nitro-burning machines down the quarter-mile.

Index

More Great Titles From CarTech®

FUNNY CAR FEVER by Steve Reyes. There wasn't always a class for these "funny-looking" cars. In the mid-1960s, many of drag racing's fastest drivers were outgrowing the Super Stock and Factory Experimental classes, building cars that stretched and eventually broke the rules. Promoters discovered they could pair up these altered-wheelbase, injected, blown machines in exhibition match races—and the spectators came running. Rivalries were born, the Funny Car class was created, and the cars kept getting faster and faster. Funny Car Fever is a humorous, heart-felt, first-hand account of the most exciting and memorable years of the Funny Car class. Steve Reyes has included over 350 of his favorite images and more than a few never-before-heard stories to bring the feeling of the class and the era home to you. Softbound, 9 x 9 inches, 192 pages, 144 color & 220 b/w photos. ***Part # CT443***

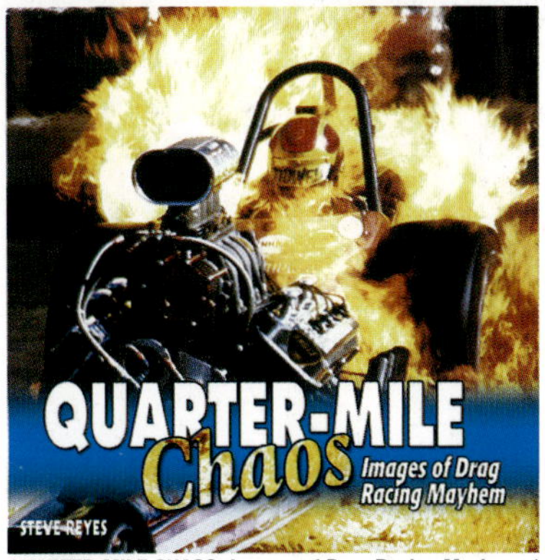

QUARTER-MILE CHAOS: Images of Drag Racing Mayhem by Steve Reyes. Quarter-Mile Chaos looks at the treacherous side of drag racing's golden age, with almost 200 rare & stunning photos from the late '60s & early to mid 1970s capturing terrifying fires, explosions, and crashes, all byproducts of the quest to go faster. Armed with just a couple cameras and some film, veteran drag racing photographer Steve Reyes roamed the nation's hazardous strips in search of the perfect action photo. The result is some of the most breathtaking drag racing imagery ever recorded. Hardbound, 10 x 10 inches, 172 pages, 180 b/w & 20 color photos. ***Part # CT425***

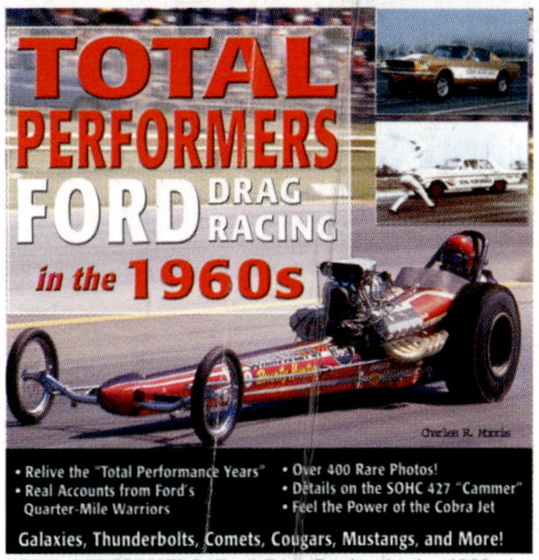

TOTAL PERFORMERS: Ford Drag Racing in the 1960s by Charles Morris. The 1960s was arguably the most important decade for drag racing. It had exciting cars, thrilling races, and, most importantly, factory participation. Among the best high-performance cars and engines were those coming from Dearborn, Michigan. Ford Motor Company's "Total Performance Years" saw a breakthrough as drag racing helped the younger, performance- and style-conscious consumer to begin receiving some recognition. Accomplished journalist and life-long Ford aficionado Charles Morris takes you back to the Total Performance Years through first-hand accounts as well as 400+ rare photos. Hardbound, 10 x 10 inches, 192 pages, 130 color & 290 b/w photos. ***Part # CT407***

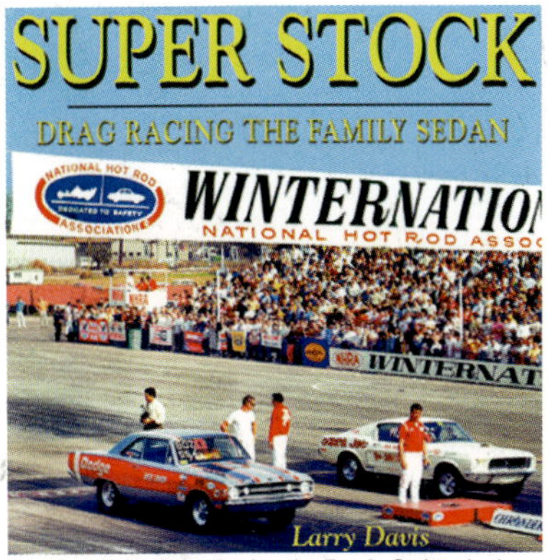

SUPER STOCK: Drag Racing the Famiy Sedan by Larry Davis.Super Stock takes a look at what was, in the 1960s, the most popular class of drag racing—factory Super Stock. It traces the evolution of the cars, the engines, the rules, the personalities, and many of the teams, from its beginnings in the mid 1950s through to the 1960s and the era of the Super Stock 409s, Ramchargers, 421 Pontiacs, and 406 Fords. Includes first-person accounts of what drag racing was really like in the early 1960s. Appendices include all of the major event winners and the rules defining the classes as well as information detailing the engines and chassis competing in Top Stock categories. Hardbound, 10 x 10 inches, 212 pages, 300 b/w & 10 color photos. ***Part # CT953***

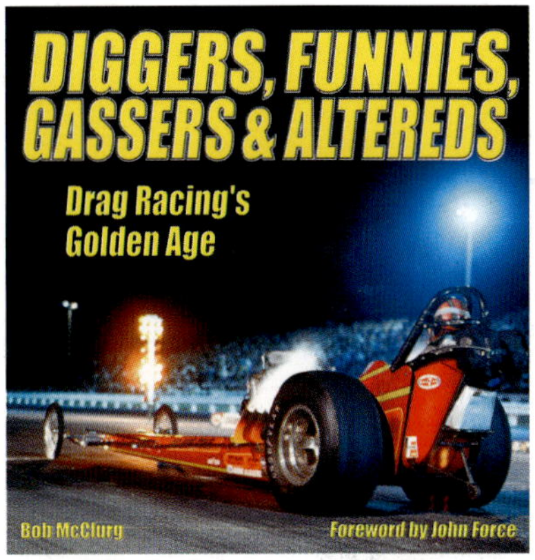

DIGGERS, FUNNIES, GASSERS, & ALTEREDS by Bob McClurg. In the '60s, drag racing evolved from a grassroots effort to a full-blown professional motorsport—along the way, it created some of the most exciting racing & race cars ever built. Bob McClurg, an accomplished magazine writer & photographer, is best known for his drag racing images of the '60s and '70s—his lens captured the Roadsters, Gassers, Altereds, Top Fuel, Funny Cars, Pro/Stocks, & even the modern age of nostalgia drag racing. With more than 350 color and black-and-white photos, this book is an exciting visual history of the Golden Age of drag racing. Hardbound, 10 x 10 inches, 192 pages, 150 b/w photos, 200 color photos. ***Part # CT990***

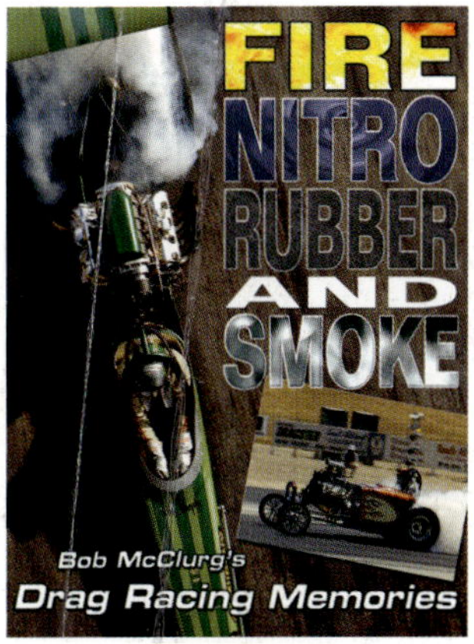

FIRE, NITRO, RUBBER, AND SMOKE: Bob McClurg's Drag Racing Memories by Bob McClurg. One of the nation's leading drag racing photojournalists, Bob McClurg captured drag racing's most fondly remembered era from the starting line to the finish line. Through extensive captions he tells the stories of the cars, the racers, and the events that made drag racing such a great sport. He includes everything from both front- and rear-engine dragsters to funny cars, gassers, altereds, pro stocks, and even exhibition cars. The sights, smells, and sounds of the golden age of drag racing come to life through these photos and provide a guaranteed trip down memory lane for any drag racing fan. Hardbound, 8-1/2 x 11 inches, 144 pages, approx. 300 color photos. ***Part # CT437***

CarTech®,Inc.
39966 Grand Avenue, North Branch, MN 55056
Telephone: (651) 277-1200 or (800) 551-4754 Fax: (651) 277-1203
www.cartechbooks.com